Arsenic & Rice

Andrew A. Meharg • Fang-Jie Zhao

Arsenic & Rice

Andrew A. Meharg
University of Aberdeen
Aberdeen
United Kingdom

Fang-Jie Zhao
Rothamsted Research
Harpenden
United Kingdom and
Nanjing Agricultural University
Nanjing
China

ISBN 978-94-007-2946-9 e-ISBN 978-94-007-2947-6
DOI 10.1007/978-94-007-2947-6
Springer Dordrecht Heidelberg London New York

Library of Congress Control Number: 2012930919

Printed on acid-free paper

Springer is part of Springer Science+Business Media (www.springer.com)

Acknowledgements

The work synthesised in this book is indebted to many people who have actively been involved in the research presented. Key amongst these are colleagues at Aberdeen (Joneil Abedin, Eureka Adomako, Anne-Marie Carey, Claire Deacon, Joerg Feldmann, Helle Hansen, Daniel Lou-Hing, Adrien Mestrot, Meher Niger, Gareth Norton, Adam Price, Andrea Raab, Mahmud Sumon and Paul Williams) and Rothamsted (Sarah Dunham, Asaduzzaman Khan, Renying Li, Wenju Liu, Charlotte Lomax, Steve McGrath, Jackie Stroud, Yuhong Su, Zhongchang Wu, Xiaoyan Xu and Wenling Ye) with whom we have worked with over the years, as well as international colleagues based in Bangladesh (Shahid Hossain, Rafiql Islam, Mazibur Rahman, ARM Solaiman), Australia (Enzo Lombi), China (Ying Lu, Ping Wu, Yongguan Zhu), USA (Tracy Punshon, Kirk Schekel), Spain (Angel Carbonell-Barrachina, Toni Signes-Pastor, Antia Villada), Japan (Jian Feng Ma), India (Tapash Dasgupta) and UK (Hugh Brammer, Claudia Cassini, Bill Davison, Parvez Haris, Katie Moore, Dave Polya and Hao Zhang) who have also been central to this work.

We are grateful to the following people who have kindly provided images, graphs or data that are used in the book: Katja Bogdan, Ann-Marie Carey, John Duxbury, Scott Fendorf, Wenju Liu, Enzo Lombi, Jian Feng Ma, Adrien Mestrot, Katie Moore, Lenny Winkel and Maozhong Zheng. We thank Steve McGrath for commenting on the last four chapters of the book.

Thanks are due to our families Caroline, Annie, Christine and Kenny (AAM) and Xiaoyun and Tina (FJZ) for their support over the years of writing this book.

Contents

Abbreviations

As	Arsenic
As(III)	Arsenite
As(V)	Arsenate
CRM	Certified reference material
DMA	Dimethylarsinic acid
DMA(III)	DMA with trivalent arsenic
DOM	Dissolved organic matter
EFSA	European Food Safety Authority
FAO	Food and Agriculture Organization
GF-AAS	Graphite furnace – atomic absorption spectroscopy
GSH	Glutathionine
HG-AAS	Hydride generation – atomic absorption spectroscopy
HG-AFS	Hydride generation – atomic fluorescence spectroscopy
HPLC	High-performance liquid chromatography
ICP-MS	Inductively-coupled plasma mass spectrometry
ICP-OES	Inductively-coupled optical emission spectroscopy
LOD	Limit of detection
MMA	Monomethylarsonic acid
MMA(III)	MMA with trivalent arsenic
NAA	Neutron activation analysis
NIP	Nodulin 26-like intrinsic protein
OVT	Ovular vascular trace
PC	Phytochelatin
QTL	Quantitative trait locus
SIMS	Secondary ion mass spectrometry
TMA	Tetramethylarsonium ion
USDA	United States Department of Agriculture
US EPA	United States Environmental Protection Agency
WHO	World Health Organization
XANES	X-ray absorption near edge structure
XAS	X-ray absorption spectrometry
XRF	X-ray fluorescence

Chapter 1
Introduction

1.1 Arsenic Exposure from Rice

The first solid food that most humans eat as weaning babies is rice, because of its blandness, lack of allergen reactions and material properties that give rise to a palatable porridge (Meharg et al. 2008). It is also the dietary staple for half the world's population (Meharg et al. 2009). Rice is approximately tenfold elevated in arsenic concentration compared to all other dietary grain staples (Williams et al. 2007a, b). The major component species of total arsenic in rice grain is inorganic arsenic (arsenate and arsenite), a class 1, non-threshold carcinogen (Meharg et al. 2009). Inorganic arsenic gives rise to a range of cancers: lung bladder and skin being the most prominent (NRC 2001; WHO 2004). Chronic exposure to inorganic arsenic species is also implicated in a range of other negative health impacts such as hypertension, diabetes, and premature births (NRC 2001; WHO 2004).

While there is rightful concern regarding high levels of inorganic arsenic exposure to ~100 million people around the world through elevated drinking water supplies (Ravenscroft et al. 2009; Smedley and Kinniburgh 2002), arsenic from rice is the largest dietary source of arsenic to the world's population with no elevated arsenic in their drinking water (EFSA 2009; Meacher et al. 2002; Meharg et al. 2009; Meliker et al. 2006; Tsuji et al. 2007; Yost et al. 2004). Even in many of the countries with highly elevated arsenic in drinking waters, because those countries at the heart of the arsenic drinking water crisis in southeast (SE) Asia and the Indian subcontinent have subsistence rice diets, rice is still a major dietary contributor to arsenic intake, and indeed may be the dominant source, particularly when drinking water sources have been reduced through mitigation (Kile et al. 2007; Mondal and Polya 2008; Ohno et al. 2007).

Even at baseline (i.e. not further elevated through anthropogenic activity), arsenic in rice is problematic (Lu et al. 2010; Meharg et al. 2009); at its worst it is predicted that 22 in 10,000 Bangladeshi population will suffer bladder and lung cancers from lifetime exposures to "natural" levels of arsenic in rice (Meharg et al. 2009). If rice is grown on geogenically, naturally, arsenic enriched soils, rice arsenic level

A.A. Meharg and F.J. Zhao, *Arsenic & Rice*, DOI 10.1007/978-94-007-2947-6_1,

further elevated from average baseline may be expected (Lu et al. 2010). Anthropogenic elevation of arsenic in rice grain occurs from three major pollution scenarios:

(a) Irrigation of rice paddies with groundwater elevated in arsenic, as occurs in Bangladesh and West Bengal India (Duxbury et al. 2003; Meharg and Rahaman 2003; Pal et al. 2009; Williams et al. 2006);
(b) Contamination of paddy soils from industrial and mining activity, with this problem being extensive over SE Asia (Liao et al. 2005; Williams et al. 2009; Zhu et al. 2008);
(c) Growing paddy rice on soil previously treated with arsenical pesticides, as occurs in South Central USA (Williams et al. 2007).

In the areas of the Indian sub-continent impacted by groundwater arsenic, not only does the irrigation with contaminated groundwater lead to elevation of arsenic in rice, rice also effectively scavenges arsenic from its cooking water (Bae et al. 2002; Pal et al. 2009) further elevating dietary exposure, remembering that the populace affected by this elevated arsenic in rice are also exposed to elevated arsenic in drinking water. This entwining of groundwater and rice exposure pathways creates many logistical problems with respect to the management of water resources in the affected regions (Meharg and Raab 2010).

As rice is traded locally, nationally and internationally, rice elevated in arsenic in one region may become the food staple of populations geographical remote from the food production source (Meharg et al. 2009; Meharg and Raab 2010). This makes arsenic in rice a trans-boundary concern of global consequence. The European Food Safety Authority (EFSA 2009) has recently evaluated dietary sources of arsenic to the European populace, and concluded that grain staples contributed ~60% of dietary exposure to inorganic arsenic, the species of most concern (EFSA 2009), with rice dominating this grain exposure as rice has typically ~10-fold higher concentration of inorganic arsenic in grain than other crops such as wheat or barley. To put these exposures into context, tap water contributed <5% inorganic arsenic to the European diet. Average rice consumption in the United Kingdom (UK), for example, is only 10 g/day (Meharg 2007), which is not untypical for a Western European diet (Meharg et al. 2009). For specific subgroups in Europe, such as those who follow SE Asian and Indian sub-continent dietary patterns, exposure to inorganic arsenic from rice is much higher. On average a UK Bangladeshi consumes 250 g of rice per day, with the Bangladeshi community in the UK constituting 5% of the population (Meharg 2007). The population of countries such as Bangladesh, Laos and Myanmar typically consume 400–500 g of rice per day (Meharg et al. 2009).

Besides rice subsistence diets, rice is the mainstay of restricted diets such as vegan, macrobiotic and dietary item avoidance regimens, due to the origin of these diets on Eastern cuisine (vegan and macrobiotic) and to its low gluten (wheat intolerance). For breast cancer patients avoiding animal milks (due to their hormone content) and for lactose intolerance patients, rice milk may be consumed to replace animal milks in the diet (Kushi 2004).

1.2 Historical Context

Initial studies regarding arsenic concentration of rice grain were first published for Taiwan (Schoof et al. 1998), the US (Schoof et al. 1999; Tao and Bolger 1998), and Vietnam (Phuong et al. 1999), identifying rice as high in total and inorganic arsenic, stating that it may be an important dietary input. These formative studies had little context in which to place their findings and did not show whether the rice they analysed was anthropogenically contaminated or not. They also could not extrapolate their findings more generically because of the limited number of samples analysed.

The role of plant and soil factors responsible for arsenic accumulation in rice started to be unravelled in the first decade of the twentieth century. Abedin et al. (2002a, b) identified that the irrigation of paddy rice with arsenic elevated water to the levels commonly found in the groundwater in Bangladesh and West Bengal, India, may be of concern, placing their physiological studies into arsenic assimilation by rice into this context. This was shortly followed by the first field surveys of arsenic in rice, identifying that there was indeed extensive arsenic contamination of rice, and paddy soil, in Bangladesh (Duxbury et al. 2003; Meharg and Rahman 2003).

These findings provided impetus to further study, resulting in the first paper to place arsenic in rice in a global context, leading to a realization that EU, US and Bangladeshi rice was elevated above "natural", and the first to realise that arsenic speciation in rice varied between different rice producing regions (Williams et al. 2005). The findings for Bangladesh were further clarified by a detailed rice grain survey (Williams et al. 2006), while concerns regarding US (Williams et al. 2007a), EU (Williams et al. 2007b) and Chinese (Zhu et al. 2008) rice were also characterized. Note that rice is widely exported, globalizing problems regarding arsenic in rice from specific elevated locations (Meharg et al. 2009; Williams et al. 2005; Zavala and Duxbury 2008). The most detailed global assessment of total and inorganic arsenic concentration of rice grain to date was published by Meharg et al. (2009), enabling potential cancer risks from rice to be calculated on a regional basis. This study shows an elevated risk of bladder and lung cancers from rice, based on the most up to date US Environmental Protection Agency (EPA) modelling of inorganic arsenic cancer risks, and that those risks are highest for countries such as Bangladesh that have very high rice consumption rates and highly contaminated rice from anthropogenic activity.

1.3 Biogeochemistry of Paddy Soils

Arsenic is problematic in rice due to the fact that rice is the only major crop grown anaerobically (i.e. under flooded conditions), and that rice is particularly efficient at assimilating some forms of arsenic, particularly those generated under anaerobic conditions, and exporting them to grain (Williams et al. 2007; Xu et al. 2008).

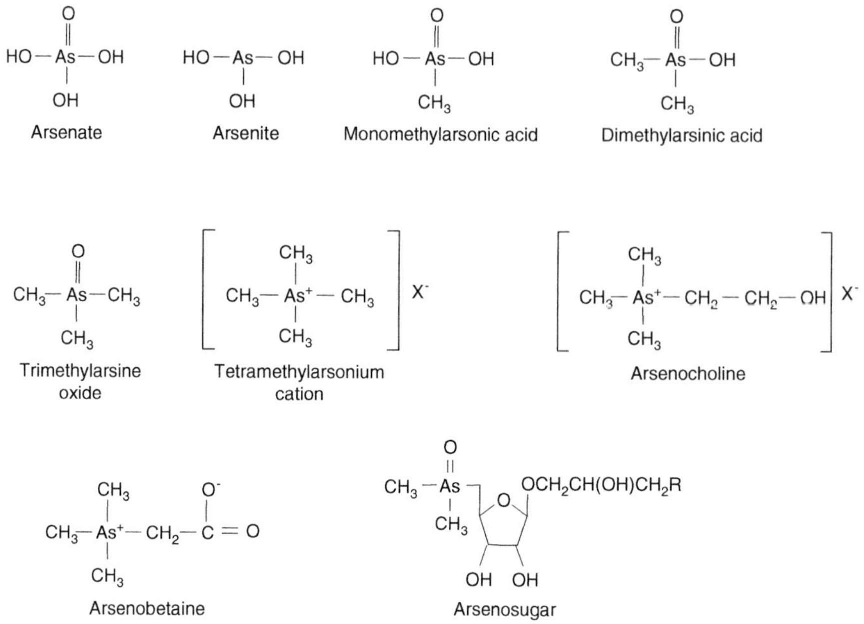

Arsenate

Arsenite

Monomethylarsonic acid

Dimethylarsinic acid

Trimethylarsine oxide

Tetramethylarsonium cation

Arsenocholine

Arsenobetaine

Arsenosugar

Fig. 1.1 Structural formulae of arsenic species

The element arsenic exists in a multitude of different chemical species in biological tissues, soils, waters and minerals, many of which are biotically and abiotically inter-convertible under a range of conditions observed in terrestrial and marine environments (Cullen and Reimer 1989). A list of the "free", that is not ligand co-ordinated, arsenic species routinely observed are given in Fig. 1.1. These can be considered as inorganic (arsenate and arsenite) or organic (including monomethylarsonic acid [MMA], dimethylarsinic acid [DMA], tetramethylarsonium [TMA], asenobetaine (AB) and arsenosugars). The inorganic species are generally more acutely toxic than organic species (Aposhian et al. 2004), with the exception of trivalent MMA(III) and DMA(III), which are intermediates of the arsenic methylation pathway, and organic species developed for chemical warfare, but these chemical agents are not found naturally, and only exist in nature highly localized around a small number of munition manufacture and testing sites (Arao et al. 2009; Baba et al. 2008).

The inorganic species arsenate [As(V)] and arsenite [As(III)] are redox sensitive, arsenite predominating under reduced and arsenate under oxidized conditions (Zhao et al. 2010). This interchange of species can be driven chemically through changing Eh and pH, as well as the presence of chemical oxidants and reductants, or enzymatically. Arsenate reductases, which are widespread in biota, can reduce arsenate to arsenite. Arsenate can be used as a terminal electron acceptor (Heimann et al. 2007), while arsenite may be oxidized by certain microbes to produce energy (lithotrophy) (Rhine et al. 2006). Arsenite can be methylated aerobically or anaerobically

(Cullen and Reimer 1989). In soils methylated arsenic species can be either partially demethylated or totally mineralized (Gao and Burau 1997; Huang et al. 2007).

With respect to plant uptake and transport, protonated arsenic species (arsenite, MMA and DMA) can behave like silicic acid analogues (Li et al. 2009a, b) and arsenate, and potentially deprotonated DMA, as phosphate analogues (Karim et al. 2009). These species have varying affinities for minerals present in the soil. Under oxidized conditions arsenate has a high affinity for iron oxyhydroxides (FeOOH) and manganese oxides (Chen et al. 2005), which makes it relatively immobile in soils, while arsenite has a lower affinity for these solid phases, making it more mobile. Under strongly reduced conditions arsenic can be precipitated as sulphide minerals such as arsenopyrite (Smedley and Kinniburgh 2002). The humic and fulvic acids that constitute dissolved organic matter in soil pore waters compete with arsenate for anion exchange sites.

Arsenic speciation is highly dynamic over the range of redox potential found in paddy fields, and those redox conditions vary spatially and temporally throughout the growing season (Dittmar et al. 2007; Takahashi et al. 2004). The flooding regimen is an obvious driver for redox, as is the vertical gradient with atmospheric oxygen perfusing down the soil profile. Rice roots aerate their rhizosphere to enable roots to survive in reduced conditions, creating redox gradients from the root surface to the bulk soil, leading to the formation of iron plaque on the root surface and in the rhizosphere (Chen et al. 2005; Liu et al. 2006).

1.4 Plant Physiology

The complexity of arsenic dynamics in soil is mirrored by that in the rice plant. Plant roots assimilate arsenic species through both silicic acid pathways (arsenite, protonated MMA and DMA) (Ma et al. 2008; Li et al. 2009a) and through phosphate transport pathways (Abedin et al. 2002b; Wu et al. 2011). Phosphate is an essential, and usually limiting, macronutrient, while rice is a silicon accumulator, thus rice is efficient at acquiring silicic acid and phosphate from the soil, making it efficient at assimilating arsenic analogues of these moieties. It is this efficiency at silicic acid/arsenite assimilation, combined with the mobilization of arsenite under reduced conditions that sets rice apart with respect to high grain arsenic burdens, as compared to crops grown under aerobic soil conditions (Zhao et al. 2010).

Once within the plant the arsenic species undergo metabolism, complexation, symplastic transport, sub-cellular localization, xylem transport to shoots and grain, with potential remobilization from shoot to grain via phloem during grain fill. Unravelling the molecular regulation of these processes is complicated as the arsenic species also exert toxicological action through inhibition of ATP formation and other phosphorylation processes, oxidative stress and binding to protein sulphylhydryl groups amongst others (Meharg and Hartley-Whitaker 2002). This toxicological action leads to grain yield reduction, further exacerbating agronomic concerns regarding arsenic in paddy rice cultivation (Panaullah et al. 2009).

While the path of arsenic from soil minerals to rice grain is dynamic and complex, arsenic speciation in grain is dominated by inorganic arsenic and DMA (Williams et al. 2005). The relative proportions of these two species in grain vary considerably between rice growing regions. It has been assumed that DMA may be produced *in planta* from methylation of inorganic arsenic, but recent studies have cast doubt on this possibility. Alternatively, DMA may be directly assimilated from soil. Both environmental factors and plant genetics also play a role in relative grain proportions of these two arsenic species (Williams et al. 2005).

The chronic toxicology of the organic arsenic species is still somewhat uncertain due to lack of knowledge regarding how they are metabolised in the human body, but, besides DMA, organic arsenic species have limited relevance to human arsenic exposure from rice or other grain staples, or indeed from water supplies, as other organic species are predominantly of marine origin (Francesconi 2007). While risk assessment of arsenic from rice focuses primarily on inorganic arsenic species, the DMA component should not be ignored.

1.5 Summary

The first papers to identify that rice may be an important dietary source of arsenic were only published from 1998, 13 years before the writing of this book. Our knowledge of the scale of the problem has grown immensely since then, enabling rice to be placed in the context of other dietary sources (EFSA 2009). At the same time the soil biogeochemistry and plant physiology leading to high arsenic levels in rice grain has been explored, making much progress. Strategies for reducing arsenic in rice grain are essential. This book synthesises knowledge regarding arsenic in rice, focussing on human exposure from arsenic in rice and how this may be mitigated through manipulation of plant genetics and physiology, and paddy field management. In addition, the information about arsenic concentration and speciation in other food crops is also given.

References

Abedin J, Cresser M, Meharg AA, Feldmann J, Cotter-Howells J (2002a) Arsenic accumulation and metabolism in rice (*Oryza sativa* L.). Environ Sci Technol 36:962–968

Abedin MJ, Feldmann J, Meharg AA (2002b) Uptake kinetics of arsenic species in rice (*Oryza sativa* L.) plants. Plant Physiol 128:1120–1128

Aposhian HV, Zakharyan RA, Avram MD, Sampayo-Reyes A, Wollenberg ML (2004) A review of the enzymology of arsenic metabolism and a new potential role of hydrogen peroxide in the detoxification of the trivalent arsenic species. Toxicol Appl Pharmacol 98:327–335

Arao T, Maejima Y, Koji B (2009) Uptake of aromatic arsenicals from soil contaminated with diphenylarsinic acid by rice. Environ Sci Technol 43:1097–1101

Baba K, Arao T, Maejima Y, Watanabe E, Eun H, Ishizaka M (2008) Arsenic speciation in rice and soil containing related compounds of chemical warfare agents. Anal Chem 80:5768–5775

Bae M, Watanabe C, Inaoka T, Sekiyama M, Sudo N, Bokul MH, Ohtsuka R (2002) Arsenic in cooked rice in Bangladesh. Lancet 360:1839–1840
Chen Z, Zhu YG, Liu WJ, Meharg AA (2005) Direct evidence showing the effect of root surface iron plaque on arsenite and arsenate uptake into rice (*Oryza sativa*) roots. New Phytol 165:91–97
Cullen WR, Reimer KJ (1989) Arsenic speciation in the environment. Chem Rev 89:713–764
Dittmar J, Voegelin A, Roberts LC, Hug SJ, Saha GC, Ali MA, Badruzzaman ABM, Kretzschmar R (2007) Spatial distribution and temporal variability of arsenic in irrigated rice fields in Bangladesh. 2. Paddy soil. Environ Sci Technol 41:5967–5972
Duxbury JM, Mayer AB, Lauren JG, Hassan N (2003) Food chain aspects of arsenic contamination in Bangladesh: effects on quality and productivity of rice. J Environ Sci Health A Tox Hazard Subst Environ Eng 38:61–69
EFSA (2009) Panel on Contaminants in the Food Chain (CONTAM)/Scientific opinion on arsenic in food. EFSA J 7:1351
Francesconi KA (2007) Toxic metal species and food-regulations – making a healthy choice. Analyst 132:17–20
Gao S, Burau RG (1997) Environmental factors affecting rates of arsine evolution from and mineralization of arsenicals in soil. J Environ Qual 26:753–763
Heimann AC, Blodau C, Postma D, Larsen F, Viet PH, Nhan PQ, Jessen S, Duc MT, Hue NTM, Jakobsen R (2007) Hydrogen thresholds and steady-state concentrations associated with microbial arsenate respiration. Environ Sci Technol 41:2311–2317
Huang JH, Scherr F, Matzner E (2007) Demethylation of dimethylarsinic acid and arsenobetaine in different organic soils. Water Air Soil Pollut 182:31–41
Karim A, Raab A, Feldmann J, Ghaderian SM, Meharg AA (2009) An arsenic accumulating, hyper-tolerant brassica, *Isatis capadocica* Desv. New Phytol 184:41–47
Kile ML, Houseman EA, Breton CV, Smith T, Quamruzzaman O et al (2007) Dietary arsenic exposure in Bangladesh. Environ Health Perspect 115:889–893
Kushi M (2004) The Macrobiotic way, 3rd edn. Avery, New York
Li RY, Ago Y, Liu WJ, Mitani N, Feldmann J, McGrath SP, Ma JF, Zhao FJ (2009a) The rice aquaporin Lsi1 mediates uptake of methylated arsenic species. Plant Physiol 150:2071–2080
Li RY, Stroud JL, Ma JF, McGrath SP, Zhao FJ (2009b) Mitigation of arsenic accumulation in rice with water management and silicon fertilization. Environ Sci Technol 43:3778–3783
Liao X-Y, Chen T-B, Liu Y-R (2005) Soil As contamination and its risk assessment in areas near the industrial districts of Chenzhou City, Southern China. Environ Int 2005:791–798
Liu WJ, Zhu YG, Hu Y, Williams PN, Gault AG et al (2006) Arsenic sequestration in iron plaque, its accumulation and speciation in mature rice plants (*Oryza sativa* L.). Environ Sci Technol 40:5730–5736
Lu Y, Dong F, Deacon C, Gan H, Zhao L, Raab A, Meharg AA (2010) Arsenic accumulation and phosphorus status in two rice (*Oryza sativa* L.) cultivars surveyed from fields in South China. Environ Pollut 158:1536–1541
Ma JF, Yamaji N, Mitani N, Xu XY, Su YH, McGrath SP, Zhao FJ (2008) Transporters of arsenite in rice and their role in arsenic accumulation in rice grain. Proc Natl Acad Sci USA 105:9931–9935
Meacher DM, Menzel DB, Dillencourt MD, Bic LF, Schoof RA, Yost LJ, Eickhoff JC, Farr CH (2002) Estimation of multimedia inorganic arsenic intake in the US population. Hum Ecol Risk Assess 8:1697–1721
Meharg AA (2007) Arsenic in rice – a literature review. Food Standards Agency, contract C101045
Meharg AA, Hartley-Whitaker J (2002) Arsenic uptake and metabolism in arsenic resistant and non-resistant plant species. New Phytol 154:29–43
Meharg AA, Raab A (2010) Getting to the bottom of arsenic standards and guidelines. Environ Sci Technol 44:4395–4399
Meharg AA, Rahman Md M (2003) Arsenic contamination of Bangladesh paddy field soils: implications for rice contribution to arsenic consumption. Environ Sci Technol 37:229–234

Meharg AA, Sun G, Williams PN, Adamako E, Deacon C, Zhu YG, Feldmann J, Raab A (2008) Inorganic arsenic levels in baby rice are of concern. Environ Pollut 152:746–749

Meharg AA, Williams PN, Adamako E, Lawgali YY, Deacon C, Villada A, Cambell RCJ, Sun GX, Zhu YG, Feldmann J, Raab A, Zhao FJ, Islam R, Hossain S, Yanai J (2009) Geographical variation in total and inorganic arsenic content of polished (white) rice. Environ Sci Technol 43:1612–1617

Meliker JR, Franzblau A, Slotnick MJ, Nriagu O (2006) Major contributors to inorganic arsenic intake in southeastern Michigan. Int J Hyg Environ Health 209:399–411

Mondal D, Polya DA (2008) Rice is a major exposure route for arsenic in Chakdaha block, Nadia district, West Bengal, India: a probabilistic risk assessment. Appl Geochem 23:2987–2998

National Research Council (2001) Arsenic in drinking water – 2001 update. National Academy Press, Washington, D.C., 2001

Ohno K, Yanase T, Matsuo Y, Kimura T, Rahman MH et al (2007) Arsenic intake via water and food by a population living in an arsenic-affected area of Bangladesh. Sci Total Environ 381:68–76

Pal A, Chowdhury UK, Mondal D, Das B, Nayak B, Ghosh A, Maity S, Chakraborti D (2009) Arsenic burden from cooked rice in the populations of arsenic affected and nonaffected areas and Kolkata City in West-Bengal, India. Environ Sci Technol 43:3349–3355

Panaullah GM, Alam T, Hossain MB, Loeppert RH, Lauren JG et al (2009) Arsenic toxicity to rice (*Oryza sativa* L.) in Bangladesh. Plant Soil 317:31–39

Phuong TD, Van Chuong P, Khiem DT, Kokot S (1999) Elemental content of Vietnamese rice-Part 1. Sampling, analysis and comparison with previous studies. Analyst 124:553–560

Ravenscroft P, Brammer H, Richards K (2009) Arsenic pollution, a global synthesis, RGS-IBG book series. Wiley-Blackwell, Oxford

Rhine ED, Phelps CD, Young LY (2006) Anaerobic arsenite oxidation by novel denitrifying isolates. Environ Microbiol 8:899–908

Schoof RA, Yost LJ, Eickhoff J, Crecelius EA, Irgolic K, Goessler W, Guo HR, Greene H (1998) Dietary arsenic intake in Taiwanese districts with elevated arsenic in drinking water. Hum Ecol Risk Assess 4:117

Schoof RA, Yost LJ, Eickhoff J, Crecelius EA, Cragin DW et al (1999) A market basket survey of inorganic arsenic in food. Food Chem Toxicol 37:839–846

Smedley PL, Kinniburgh DG (2002) A review of the source, behaviour and distribution of arsenic in natural waters. Appl Geochem 17:517–568

Takahashi Y, Minamikawa R, Hattori KH, Kurishima K, Kihou N, Yuita K (2004) Arsenic behavior in paddy fields during the cycle of flooded and non-flooded periods. Environ Sci Technol 38:1038–1044

Tao SSH, Bolger PM (1998) Dietary arsenic intakes in the United States: FDA total diet study, September 1991-December 1996. Food Addit Contam 16:465

Tsuji JS, Yost LJ, Barraj LM, Scrafford CG, Mink PJ (2007) Use of background inorganic arsenic exposures to provide perspective on risk assessment results. Regul Toxicol Pharmacol 48:59–68

Williams PN, Price AH, Raab A, Hossain SA, Feldmann J, Meharg AA (2005) Variation in arsenic speciation and concentration in paddy rice related to dietary exposure. Environ Sci Technol 39:5531–5540

Williams PN, Islam MR, Adomako EE, Raab A, Hossain SA et al (2006) Increase in rice grain arsenic for regions of Bangladesh irrigating paddies with elevated arsenic in groundwaters. Environ Sci Technol 40:4903–4908

Williams PN, Raab A, Feldmann J, Meharg AA (2007a) Market basket survey shows elevated levels of as in South Central US processed rice compared to California: consequences for human dietary exposure. Environ Sci Technol 41:2178–2183

Williams PN, Villada A, Deacon C, Raab A, Figuerola J et al (2007b) Greatly enhanced arsenic shoot assimilation in rice leads to elevated grain levels compared to wheat and barley. Environ Sci Technol 41:6854–6859

Williams PN, Lei M, Sun G-X, Huang Q, Lu Y, Deacon C, Meharg AA, Zhu Y-G (2009) Occurrence and partitioning of cadmium, arsenic and lead in mine impacted paddy rice: Hunan, China. Environ Sci Technol 43:637–642

World Health Organization (2004) IARC, Working Group on some drinking water disinfectants and contaminants, including arsenic, vol 84. Lyon. Monograph 1

Wu ZC, Ren HY, McGrath SP, Wu P, Zhao FJ (2011) Investigating the contribution of the phosphate transport pathway to arsenic accumulation in rice. Plant Physiol 157:498–508

Xu XY, McGrath SP, Meharg A, Zhao FJ (2008) Growing rice aerobically markedly decreases arsenic accumulation. Environ Sci Technol 42:5574–5579

Yost LJ, Tao SH, Egan SK, Barraj LM, Smith KM, Tsuji JS, Lowney YW, Schoof RA, Rachman NJ (2004) Estimation of dietary intake of inorganic arsenic in US children. Hum Ecol Risk Assess 10:473–483

Zavala YJ, Duxbury JM (2008) Arsenic in rice: I. Estimating normal levels of total arsenic in rice grain. Environ Sci Technol 42:3856–3860

Zhao FJ, McGrath SP, Meharg AA (2010) Arsenic as a food-chain contaminant: mechanisms of plant uptake and metabolism and mitigation strategies. Annu Rev Plant Biol 61:7.1–7.25

Zhu Y-G, Sun G-X, Lei M, Teng M, Liu Y-X, Chen N-C, Hong W-L, Carey AM, Meharg AA, Williams PN (2008) High percentage inorganic arsenic content of mining impacted and nonimpacted Chinese rice. Environ Sci Technol 42:5008–5013

Chapter 2
Arsenic in Rice Grain

2.1 Introduction

To understand the threat posed by arsenic in rice to human food chains, the quantity and chemical species of arsenic ingested by an individual or population have to be assessed, along with gut bioavailability and the hazard associated with that arsenic. This chapter will deal with arsenic present in commercial rice products.

2.2 Quantification of Arsenic in Rice Grain

It only started to be realised that rice was a major dietary contributor to inorganic arsenic intake when analytical developments enabling low level arsenic speciation became routine (Schoof et al. 1999). The development of inductively-coupled plasma mass spectrometry (ICP-MS) as an ultra-sensitive arsenic detector, combined with High Performance Liquid Chromatography (HPLC), enabled the robust, low level quantification and qualification required to survey arsenic speciation in rice grain (Williams et al. 2005). As the relative cost has decreased, along with increased reliability, HPLC-ICP-MS has become the "gold-standard" for arsenic speciation in foods, including rice (de la Calle et al. 2011; Meharg and Raab 2010). The first comprehensive survey of inorganic arsenic in foodstuffs, using HPLC-Atomic Fluorescence Spectroscopy (AFS), Schoof et al. (1999) stated "that rice has higher inorganic arsenic concentrations than most other foods, and consequently, diets that rely heavily on rice may contain the most inorganic arsenic."

A range of detection systems can be used to quantify arsenic, and a range of separation techniques to aid in species qualification and quantification. This section will consider techniques commonly used to quantify and speciate arsenic in rice grain. The outlined techniques are also appropriate to other plant samples. The lack of Certified Reference Materials (CRM) for arsenic speciation has been an issue for setting inorganic arsenic standards in rice (de la Calle et al. 2011; EFSA 2009; Meharg and Raab 2010), and this issue will also be considered.

A.A. Meharg and F.J. Zhao, *Arsenic & Rice*, DOI 10.1007/978-94-007-2947-6_2,

Table 2.1 Strengths and weaknesses for different analytical approaches to measure total arsenic

Technique	Relative cost	Infrastructure requirement	User expertise	Instrumental LOD (μg L^{-1})	Analytical interference
ICP-MS	High	High	High	~0.1	Limited and controllable
ICP-OES	Moderate	High	High	~5	Minor
HG-AAS	Low	Moderate	Low	~1	Considerable
GF-AAS	Moderate	Moderate	Moderate	~10	Limited and controllable
HG-AFS	Low	Low	Low	~0.1	Considerable
NNA	Very high	Very high	High		Minor

2.2.1 Total Arsenic Analysis

2.2.1.1 Digestion of Grain

With the exception of Neutron Activation Analysis (NAA) which does not require sample digestion (de la Calle et al. 2011; Freitas et al. 2008), techniques that do require digestion for total arsenic the digestion procedures are straightforward. As rice grain is a relatively simple matrix, digestion of grain is easily achieved through heating in the presence of nitric acid (Williams et al. 2005). This heating is achieved either through microwave digestion or traditional heating-block techniques. The efficacy of digestion, and subsequent analysis, can be readily checked through the use of certified reference materials (CRMs, e.g. NIST 1568a rice flour). Nitric acid is also a suitable carrier for subsequent analysis. Other variants of digestion methods include the use of nitric acid and hydrogen peroxide, or nitric acid and perchloric acid, both also producing a good recovery for arsenic (Li et al. 2009; Xu et al. 2008). As the levels of arsenic in rice and other plant samples are generally low, high purity reagents and high-quality deionized water should be used to minimize the blank.

An important point to remember with digestion is that it dilutes the sample. Rice grain arsenic can be as low as 10–20 μg kg^{-1} (Table 2.1), and a typical dilution of 100 is required to convert solid grain into a digest, given that digestion acids must be diluted to run on most instruments due to viscosity and corrosion. Therefore, digested solution can have concentrations as low as 0.1 μg L^{-1}, and selection of analytical instrumentation needs to take this into mind.

2.2.1.2 Analysis of Total Arsenic

ICP-MS, ICP-Optical Emission spectroscopy (ICP-OES), Hydride Generation – Atomic Absorption Spectroscopy (HG-AAS), Graphite Furnace – AAS (GF-AAS), Hydride Generation – Atomic Fluorescence Spectroscopy (HG-AFS) and NAA can all be used to detect total arsenic; all have strengths and weakness, which are outlined in Table 2.1.

There is considerable variation in cost, infrastructure required to support analysis, limit of detection (LOD) and being subject to analytical interferences between the techniques. Given the LOD constraints mentioned above, only ICP-MS and HG-AFS are widely suitable for measurement of total arsenic in grain, though if higher grain arsenic samples are expected (>100 μg kg^{-1} grain arsenic on a dry weight basis), ICP-OES and HG-AAS come into play.

While HG-AFS competes with ICP-MS on sensitivity, and is a lot cheaper to buy and maintain, it has one major flaw when it comes to analysing rice grain in that its arsenic signal production is species dependent, with methylated species being subject to kinetic interferences due to their slower hydride formation as compared to inorganic species. Other hydride-generation based techniques for total analysis suffer similar kinetic interferences such as hydride interfaced to ICP-OES (where it is used to increase sample delivery and lower limits of detection), ICP-MS (where it is used to enhance sample delivery and to avoid potential argon chloride interferences as $^{40}Ar^{35}Cl^{+}$ ion has the same mass as $^{75}As^{+}$) or routine HG-AAS. Unless HCl is used as a digesting agent, or as a solvent extractant (Munoz et al. 2002; Torres-Escribano et al. 2008), polyatomic $^{40}Ar^{35}Cl^{+}$ interference is not a problem with rice due to it having a low chloride content. Therefore, ICP-MS collision cell technologies are not normally deployed in rice tissue quantification as they raise LOD. Where $^{40}Ar^{35}Cl^{+}$ interference may occur, for example in arsenic speciation analysis in soil solution or in plant vegetative tissues, $^{40}Ar^{35}Cl^{+}$ can be effectively eliminated by the use of helium collision gas. The deployment of an UV oxidation cell in the hydride generator mineralizes organic arsenic species, therefore, enabling hydride techniques to be used in rice analysis with confidence (de la Calle et al. 2011), assuming that suitable LODs are achieved. de la Calle et al. (2011) reports proficiency testing for a range of instrumental techniques (ICP-MS, HG-AAS, NAA) and found them all to be reliable in characterizing the total arsenic concentration of grain.

2.2.2 *Arsenic Speciation by Chromatographic Separation*

Arsenic in rice is either speciated following destructive extraction techniques (Abedin et al. 2002; Schoof et al. 1999; Williams et al. 2005) or by using non-destructive *in situ* synchrotron X-ray absorption spectroscopy (XAS) approaches (Carey et al. 2010; Lombi et al. 2009; Meharg et al. 2008c). The advantages of chromatographic techniques are better sensitivity and the ability to accurately qualify (if authentic standards are available and if suitable chromatographic separation is achieved) and quantify individual species. Sychrotron XAS approaches are only qualitative, but are not subject to potential speciation changes that could occur during chemical extraction of a sample (Meharg et al. 2008c). However, lack of appropriate standards and problems in fitting the X-ray absorption spectra of standards to those of the samples can lead to misinterpretation, especially for species present at low concentrations (Carey et al. 2010; Lombi and Susini 2009). Sample oxidation, converting arsenite to arsenate, may also occur during XAS measurements (Feldmann et al. 2009).

2.2.2.1 Chemically Extracting Arsenic Species

Arsenic speciation in rice is dominated by inorganic arsenic (arsenate and arsenite) and DMA with trace amounts of MMA and tetramethylarsonium (TMA) (Hansen et al. 2011). For grain (as well as other rice tissues) arsenic species are conventionally extracted into dilute acids, normally nitric acid or trifluoroacetic acid, followed by heating to ~90°C on a hot-block or by microwave digestion (Abedin et al. 2002; Heitkemper et al. 2001; Norton et al. 2009; Williams et al. 2005). While methylated species remain unchanged under such conditions, arsenite is partially converted to arsenate (Abedin et al. 2002). Thus, it is conventional to report the sum of arsenate and arsenite as inorganic arsenic. As the primary interest in inorganic arsenic speciation in rice and other food crops is to calculate human ingestion, this lack of discrimination between arsenate and arsenite is not problematic as arsenite and arsenate are both readily assimilated in mammalian systems where they interconvert (Juhasz et al. 2006).

If inorganic arsenic speciation needs to be differentiated, as in physiological studies, then mild extraction procedures are required, which is also the case where the oxidation state of methylated species is to be ascertained (Li et al. 2009), or if arsenic-phytochelatin complexes are under study (Raab et al. 2004). Where oxidation state is to be determined a suitable buffer (e.g. phosphate buffer solution), degassed with nitrogen to remove oxygen, and immediate analysis following extraction, is required. For the analysis of arsenic-phytochelatin complexes, samples are extracted with an acidic solution (e.g. 1% formic acid) to stabilise the complexes, as the arsenite-thiol complexes dissociate easily under neutral or alkaline conditions (Raab et al. 2004).

When extracting arsenic from rice it is important to establish extraction efficiency. The most thorough way of doing this is to independently ascertain total concentration of arsenic in the extract and express this as a percentage of the total arsenic in the grain. Alternatively, if total arsenic determination on the extraction is not obtained then the sum of arsenic species can be used as another relevant measure. Another important consideration, when speciating arsenic through chromatography, is chromatographic column recovery; this is the sum of species expressed as a percentage of the total arsenic in the extract used for speciation. This column recovery reveals if any of the arsenic present in the speciation extract has not eluted from the column. These various aspects of arsenic extraction efficiency and column recoveries have been well established for standard speciation extraction techniques such as the widely used trifluoroacetic acid (Williams et al. 2005) and nitric acid protocols (Raab et al. 2008). They show that extraction and column recoveries vary among rice cultivars, presumably due to arsenic being present in, as yet, unidentified macro-molecules. It is assumed, however, that as inorganic arsenic and DMA are readily extractable and column elutable (under optimized conditions for speciation), these species are not those left behind by extraction or those that are retained by the column. These "non-extractable" and "non-elutable" phases remain unknown, even following synchrotron XAS *in-situ* speciation approaches (Carey et al. 2010). Thus, while all new analytical protocols should be tested for extraction efficiency and column recoveries for rice grain or other types of samples (e.g. soil and rice vegetative tissues), for those protocols that have been tested and found satisfactory, simple reporting of the sum of inorganic arsenic and of DMA is deemed appropriate.

2.2.2.2 Species Quantification

Most methods used for arsenic speciation are based on chromatographic separation (de la Calle et al. 2011). The most common exception to this is a technique based on partitioning $AsCl_3$ formed through the reaction of inorganic arsenic with concentrated HCl into chloroform, with a suitable total arsenic analysis to quantify the extracted inorganic arsenic (de la Calle et al. 2011; Munoz et al. 2002; Torres-Escribano et al. 2008). As arsenic analysis by ICP-MS is prone to $ArCl^+$ interferences, only high resolution (HR) – ICP-MS (de la Calle et al. 2011) can be used with this extraction based technique because of the very high levels of chlorine present in the extractants. A further limitation of this technique is that it does not give organic species.

More conventionally, as the major species of interest in rice grain are neutral (arsenite) or anionic (arsenate, MMA(V), DMA(V)) under standard chromatography conditions, anion exchange chromatography is widely used to separate arsenic species in rice grain, soil and other plant materials, combined with sensitive post column detection, either ICP-MS or HG-AFS (de la Calle et al. 2011). Hansen et al. (2011), however, reported trace quantities of tetramethylarsonium (TMA), which is cationic, in rice grain.

Speciation of arsenite using anion exchange column is somewhat problematic as it elutes near the solvent front. Quantifying this solvent front peak and assigning it to arsenite is often deemed suitable in speciation analysis but may be erroneous as TMA will also elute at a similar retention time, and only when chromatography is optimal can TMA and arsenite be separated (Hansen et al. 2011). Figure 2.1 illustrates a chromatogram of rice grain with detectable TMA overlaid with a chromatogram of authentic standards. To negate this problem, and as arsenite determination in rice extracts is of dubious certainty unless considerable effort is made to stop the interconvertion of arsenate and arsenite (see Sect. 3.1), arsenite can be oxidized to arsenate by the addition of hydrogen peroxide to the extract just before performing chromatographic separation (Hansen et al. 2011). Thus, the arsenate peak equates to total inorganic arsenic when such an oxidation step is introduced.

de la Calle et al. (2011) compared HPLC-ICP-MS and HCl-choloroform extraction followed by HG-AFS or HR-ICP-MS procedures, and found that speciation efficacy for inorganic arsenic was independent of detector or extraction conditions for previously well optimized procedures.

As mentioned in Sect. 2.2.1, more exotic arsenic species can be detected in plants, including rice, though not, to date, in grain, by using mild extraction procedures. In this way MMA(III) is quantifiable in rice tissue (Li et al. 2009). Furthermore, by combining mild extraction (1% formic acid) with reverse phase chromatography and combined inorganic (ICP-MS) and organic (electrospray ionization) mass spectroscopy, phytochelatin complexed arsenic species can be qualified and quantified (Raab et al. 2004).

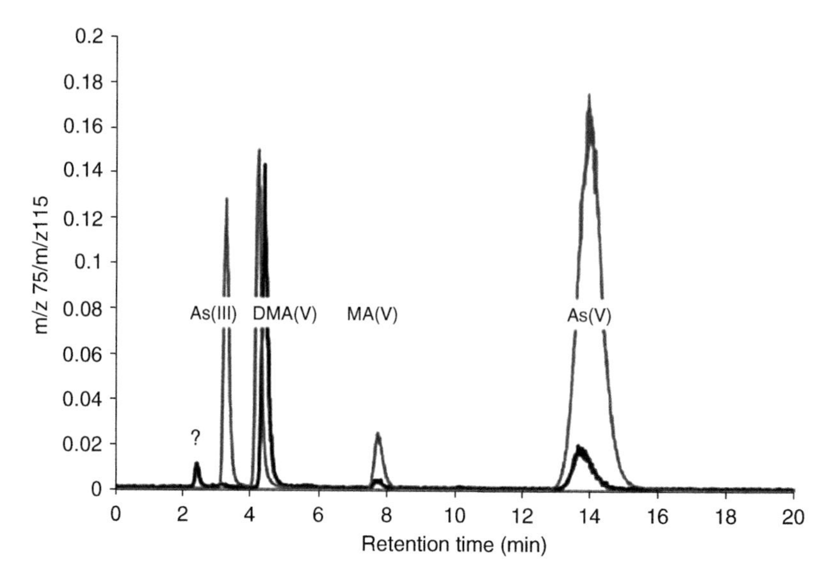

Fig. 2.1 Chromatogram of a rice grain extract (*black chromatogram*) compared with authentic standards (*grey chromatogram*) which shows the presence of an unknown peak (?) subsequently identified as TMA (Hansen et al. 2011)

2.2.3 Certified Reference Materials (CRMs)

The lack of appropriate CRMs for arsenic speciation in grain is becoming somewhat of an issue (de la Calle et al. 2011; EFSA 2009; Meharg and Raab 2010). Although scientists working within the field have circumnavigated the need for CRMs by adopting the NIST 1568a rice flour for which the total arsenic concentration is certified, as a reference for arsenic speciation analysis. Results from different laboratories are comparable (e.g. Raab et al. 2008; Williams et al. 2005; Zavala et al. 2008). de la Calle et al. (2011) recently developed a new rice flour reference material, as yet uncertified, to conduct inter laboratory studies. While having an authenticated rice flour CRM for arsenic speciation would put analysis conducted in this field onto a firmer footing, it is really for regulatory purposes that a CRM is required (de la Calle et al. 2011; EFSA 2009), and for this reason the production of such a CRM is essential for future studies in this area.

2.3 Arsenic Speciation in Rice Grain

When arsenic is chemically extracted (i.e. through the use of solvents) from grain it is found to be primarily in the form of inorganic arsenic and DMA. This dominance by inorganic arsenic is also true for other terrestrial crops where arsenic is speciated, with DMA usually being a minor component (see Chap. 8) (Meharg and Hartley-Whitaker 2002).

Rice differs from other crops in the efficiency of arsenic assimilation into the grain (Williams et al. 2007b), as well as growing in anaerobic paddy conditions which can favour arsenic methylation by soil microorganisms (see Chap. 5).

"Inorganic arsenic" is a generic term usually referring to the sum of arsenate and arsenite concentrations. Reporting individual arsenite and arsenate concentrations is normally not conducted as chemical extraction by solvents prior to analytical detection can change the relative speciation of these two entities (Abedin et al. 2002), and because toxicologically there mode of actions are interrelated, i.e. arsenate is rapidly reduced to arsenite in most cell systems (Zhao et al. 2009), including humans (NRC 2001; WHO 2004). Risk assessment of dietary arsenic exposure is conducted on inorganic arsenic, rather than arsenite or arsenate per se because of their entwined metabolism.

Rice grain can contain other arsenic species besides inorganic arsenic and DMA. Occasionally, usually dependent on having very low limits of detection for a given analytical run combined with high total grain concentrations, monomethyarsonic acid (MMA) can be observed in grain (Williams et al. 2005). As MMA is thought to be the metabolic intermediate in the widely observed metabolism of inorganic arsenic to DMA in organisms (but see Chap. 5 for an alternative methylation pathway), the presence of trace quantities MMA is not surprising. Recently TMA ion was detected in highly arsenic contaminated rice grain (Hansen et al. 2011). This compound is rarely seen in plants or soil, and may have its source in DMA that has been further methylated, possibly by soil microflora. More exotic organo-arsenic species have been found when rice is grown on soil contaminated with arsenical-based chemical agents (mainly diphenylarsinic acid) such as in some Japanese scenarios (Arao et al. 2009; Baba et al. 2008) or with the animal growth promoter Roxarsone (4-hydroxy-3-nitrobenzenearsonic acid) (Liu et al. 2009). While the chemical war agent arsenicals should be expected to be rare, Roxarsone is widely used in poultry and pig industries in some countries (such as the US and China). Where waste streams from such industries are used to manure paddy soils, then Roxarsone would have another route into the human food chain. However, many parts of the world do not use Roxarsone, and there is no detailed information for those that do that contaminated manure is ending up as paddy fertilizer.

Given that exotic (diphenylarsinic acid) and restricted (Roxarsone) arsenicals are relatively rare in most environments, and that MMA and TMA are never more than trace constituents of rice grain arsenic, grain speciation can be considered, with respect to human health concerns, to be inorganic arsenic and DMA.

The *in situ* speciation of arsenic in grain is more complex than that observed on chemical extraction, both spatially and chemically (Carey et al. 2010). For reasons discussed in Chap. 6, arsenic concentration is higher in the outer parts of grain, the parts of the grain polished off in producing white rice. Chemically extracted inorganic arsenic is dominant in this "bran" layer while DMA, given that rice varies greatly in DMA concentration, is relatively more prolific in endosperm (Sun et al. 2008). The consequence for brown versus white rice with respect to arsenic concentration and speciation of bran and endosperm are outlined in Sect. 2.4 of this chapter. If non-destructive micro X-ray absorption near edge structure (μXANES) is used to speciate arsenic *in situ*, it is found that inorganic arsenic speciation is dominated by arsenite while DMA is present at approximately equal proportion as free or sulphur complexed DMAS (Carey et al. 2010). However, whether these subtleties in arsenic speciation survive transit through human digestive tracts is another matter and, as yet, has not been ascertained. Therefore, again, arsenic speciation in rice is normally characterised by its inorganic arsenic and DMA concentrations.

This section will restrict itself, as is also true for the rest of the chapter, to commercially purchased rice. Environmental variables and agronomic conditions affecting arsenic speciation in grain are discussed in Chap. 6, here the discussion is limited to market purchased rice, i.e. those purchased for direct human consumption. Subsequent sections will deal with products made from rice. It should be noted here that subsistence rice farmers, or a populace dependent on highly localized rice mills, will experience much greater heterogeneity in supply, with the worst case scenario being farms subsisting on fields with elevated arsenic (Meharg and Raab 2010).

At present, there are many individual studies reporting arsenic speciation in rice, but these have been collected for a range of purposes (e.g. analytical development, bioavailability studies, effect of cooking or thermal treatment), and the disparity in collection and analytical procedures makes cross comparisons difficult. To date, only limited inter-country studies looking at commercially purchased rice, each internally consistent with respect to analytical procedures and sampling methodologies, have been published.

The first robust comparative study of rice available in the market place was Williams et al. (2005), which established for the first time a baseline context against which rice from different regions could be compared. They discovered that rice from different parts of the globe differed greatly in arsenic speciation, with rice produced in the US and EU having a high percentage of DMA compared to Bangladeshi and Indian rice. Meharg et al. (2009) confirmed this, extending the study to China, finding that Chinese rice was dominated by inorganic arsenic as opposed to DMA. A range of country specific surveys has added more details to our knowledge of arsenic speciation in rice grain. Table 2.2 and Fig. 2.2 summarize the studies conducted to date where rice grain has been speciated by HPLC-ICP-MS from the laboratories of the University of Aberdeen. Figure 2.2 adds in additional information from the studies of Zavala and Duxbury (2008) and Torres-Escribano et al. (2008).

It can be observed from Table 2.2 that total arsenic varies between regions. Ghana had the lowest median arsenic concentration (20 ng g^{-1}) followed by India (50 ng g^{-1}), while the USA, Italy and Thailand had over an order of magnitude greater arsenic concentrations, with China and Bangladesh being intermediate.

The inorganic arsenic concentration follows a similar trend though the fold differences between the countries decrease, and the relative positions change (Table 2.2). China has a higher percentage of inorganic arsenic than the norm, while US has the lowest inorganic arsenic percentage. This switches US inorganic arsenic concentration to intermediate and Chinese to high in this meta-analysis ranking. DMA concentration was also variable, with the US having the highest percentage of DMA and Ghana and India having the lowest. A residue of grain arsenic is not extractable, and thus inorganic arsenic and DMA do not add up to 100% of the total arsenic (Table 2.2).

Table 2.2 Arsenic speciation in market rice collected from different regions of the world

Country		N	Total As (ng g^{-1})	Inorganic As (ng g^{-1})	DMA (ng g^{-1})	% inorganic As	% DMA
Bangladesh	White	15	100 (30–300)	70 (10–210)	83 (0–50)	60 (42–86)	15 (1–49)
China	White	21	160 (106–586)	123 (70–379)	34 (15–147)	74 (22–94)	20 (11–62)
Ghana	White	7	20 (5–120)	21 (5–83)	0 (0–40)	100 (69–140)	0 (0–33)
India	White	13	50 (30–162)	30 (16–93)	10 (0–47)	57 (28–67)	15 (4–83)
	Brown	6	105 (43–148)	71 (40–94)	9 (0–18)	66 (61–124)	9 (6–19)
Italy	White	6	211 (127–318)	130 (68–161)	57 (24–90)	54 (44–73)	29 (15–41)
	Brown	5	187 (156–230)	101 (97–136)	25 (19–50)	59 (53–68)	13 (11–24)
Thailand	White	6	211 (127–318)	130 (68–161)	57 (24–90)	54 (44–73)	30 (15–41)
USA	White	13	220 (160–400)	81 (50–110)	100 (50–260)	37 (20–54)	44 (31–65)
	Brown	3	29 (11–34)	120 (60–140)	110 (40–150)	42 (41–59)	40 (32–45)

Data are compiled from Meharg et al. (2009), Raab et al. (2008), Adamoko et al. (2011)

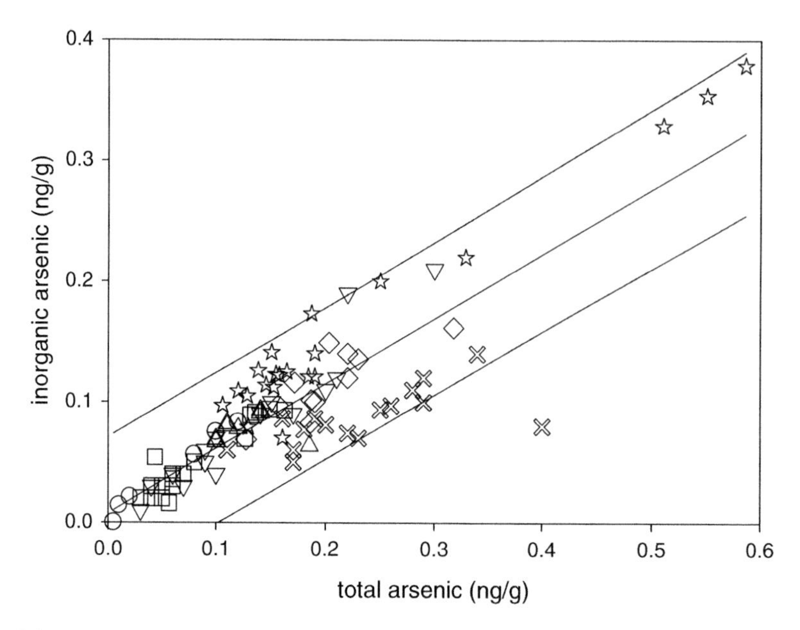

Fig. 2.2 Relationship between total arsenic and inorganic arsenic concentrations in rice grain (Data are compiled from Meharg et al. (2009), Raab et al. (2008), Adamoko et al. (2011)). *Inverted triangles* are for Bangladesh, *stars* for China, *circles* for Ghana, *squares* for India, *diamonds* for Italy, *triangles* for Thailand and *crosses* for USA. The mean regression line is shown along with the 95% confidence intervals

The variation in inorganic arsenic concentration between countries is illustrated in Fig. 2.2, which plots total arsenic versus inorganic arsenic concentration for the data summarised in Table 2.2. The plotted regression is highly significant ($P<0.001$) with an r^2 of 76.8%. Note that the regression holds best at lower concentrations. This being said, the Chinese data have a steeper slope while the US data have a shallower slope, with all other data falling on the regression line. The equation for this line is:

$$\text{Grain inorganic arsenic} = 0.01 + 0.54 \times \text{total grain arsenic}$$

Thus, on average across the dataset, inorganic arsenic concentration is 54% of total arsenic. As speciation of grain arsenic is more time consuming and technologically demanding to measure than total arsenic, the former requiring HPLC-ICP-MS, the vast majority of grain arsenic measurements are only for totals. This regression between total arsenic and inorganic arsenic can be used to calculate risk posed by rice (see Chap. 3) in the absence of detailed speciation. However, when robust speciation data are available for any given country, such as for Bangladesh, China and US (Meharg et al. 2009), then these should be used in preference.

The data examined in Table 2.2 and Fig. 2.2 are in broad agreement with directly comparable results obtained by Zavala et al. (2008) who also used HPLC-ICP-MS to speciate US rice, and with the findings of Torres-Escribano et al. (2008), Lamont (2003), Jorhem et al. (2008) who used HCl/chloroform extraction to speciate

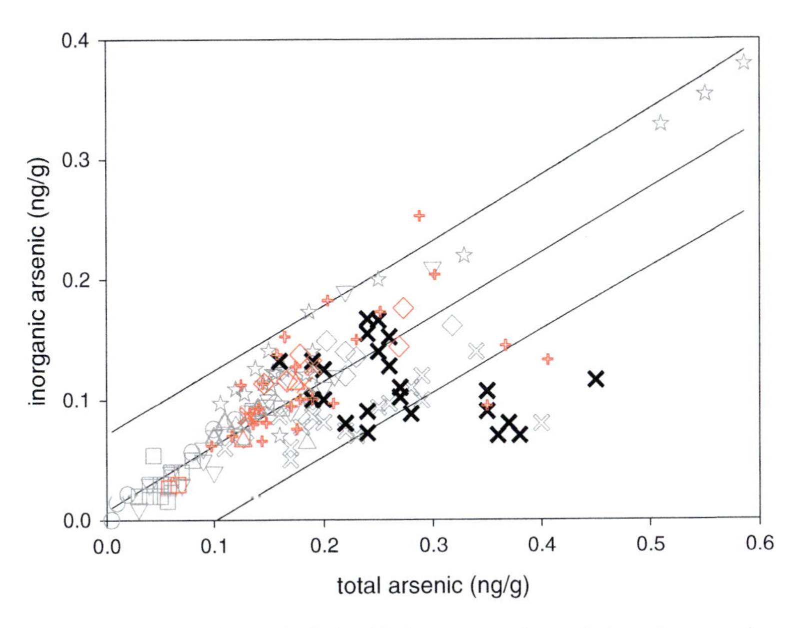

Fig. 2.3 Literature comparisons of relationship between total arsenic in grain versus inorganic arsenic concentration compared to data plotted in Fig. 2.2. *Black filled crosses* data are from Zavala et al. (2008) while the *red outlined plus* data are from Torres-Escribano et al. (2008). Data points in *grey*, along with regression lines, are those from Fig. 2.2. *Inverted triangles* are for Bangladesh, *stars* for China, *circles* for Ghana, *squares* for India, *diamonds* for Italy, *triangles* for Thailand and *crosses* for USA. The mean regression line is shown along with the 95% confidence intervals

inorganic arsenic in Spanish home grown and imported rice, US home grown rice and Swedish imported market rice, respectively. Lamont (2003) did not report total grain arsenic, so further comparison is not warranted. The other studies are plotted alongside analysis from Fig. 2.3 to provide a meta overview of the bulk of market basket surveys published to date showing how the US data reported by Zavala et al. (2008) are very similar to that produced in Fig. 2.2, while the Spanish (and other) data of Torres-Escribano et al. (2008) follow the main trend for the overall dataset. The reason why US rice has enhanced DMA (resulting in high total arsenic but relatively low inorganic arsenic concentration) is considered in Chap. 6.

2.4 Total Grain Arsenic

Total arsenic data are more widely available than speciated data as speciation is more technically demanding, more time consuming and less widely available. From Sect. 2.3, inorganic arsenic can be extrapolated from totals, and if robust speciation data are not available for a particular region, totals must be interpreted with some caution so as not to over or under estimate risks posed by the inorganic arsenic content in rice.

The widely available data of total grain arsenic enable general trends between and within countries to be considered, given the provisos outlined in the paragraph above. Figure 2.4 reports the most comprehensive market survey for total arsenic in grain published to date, representing ~900 individual measurements, from rice collected from a wide range of countries and all conducted at one laboratory, therefore, having consistent quality assurance/quality control (QA/QC) during analysis. Other regional surveys have been published (i.e. Zavala and Duxbury 2008) but these only report summary statistics and, therefore, cannot be added to the analysis in Fig. 2.4.

The total arsenic concentration of market rice from different countries of origin varies greatly with the 95th percentile of Egyptian rice, the lowest in total arsenic observed for any region to date, was below the 5th percentile for French and US rice (Fig. 2.4), reflecting also a tenfold range in total arsenic median concentrations. It is notable that industrialized nations have the highest ranked medians (France, USA, Spain, Italy and Japan) while developing countries had the lowest (Egypt, India, Nepal and Pakistan). Whether this industrialized/developing nations ranking is due to agronomic and edaphic factors or to anthropogenic pollution, or rather the balance between natural and polluting sources, is unknown as there will always be an underlying natural (geogenic) component, which probably varies considerably within a region, in any case. The sources of arsenic to rice are considered in detail in Chap. 4; in this chapter it is the consequences for the sources with respect to grain arsenic that is of interest here, but a brief summary of potential explanations of these differences follows.

The US has a long history of rice production on soil previously treated with arsenical pesticides in the major region of cotton production, the South-Central zone (Williams et al. 2007a). China has extensive mining impacted paddy soil which leads to elevation of grain arsenic, but these mining regions while large are probably only a small fraction of overall Chinese production. However, because there is specific interest in such regions, this has biased surveys with over representation of these regions (Zhu et al. 2008). Ghana has a smaller rice industry, some of it impacted by gold mining leading to elevated grain arsenic in these regions (Adomako et al. 2011). Bangladesh rice is widely grown in zones where arsenic elevated groundwater is used to irrigate rice during dry season (Boro) production, resulting in grain arsenic elevation over extensive regions of the country (Williams et al. 2006). Japanese soils are widely geogenically elevated in arsenic and Japan suffers from anthropogenic contamination of paddies (Adomako et al. 2011) but the extent of human contamination is not known. The extent to which French, Spanish, Italian and Thai rice is elevated by anthropogenic activity has not been investigated, though grain produced in these countries is elevated in arsenic compared to the Himalayan and Nile Delta regions. Lin et al. (2004) surveyed market rice (280 samples) within Taiwan and found a mean total arsenic concentration of 100 ng g^{-1}, ranging from <100 to 630 ng g^{-1}, putting Taiwan in an intermediate rank position with respect to Fig. 2.4.

An extensive survey of US rice along with a smaller numbers of samples for Spain, Italy, India, Thailand, Pakistan and Venezuelan produced rice was published by Zavala and Duxbury (2008). This study came to the same general conclusion as the data presented in Fig. 2.4, with US rice having the highest median concentrations and India and Pakistan the lowest. Although they had Bangladeshi rice at similar

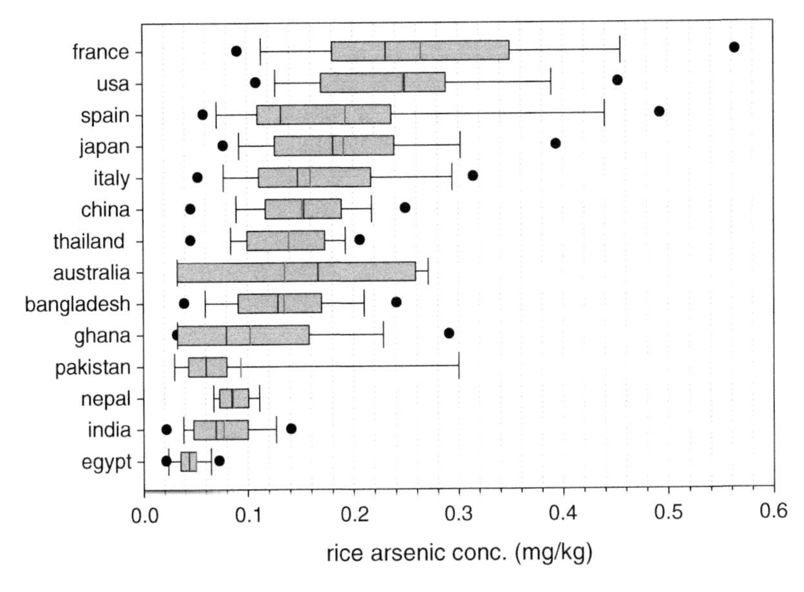

Fig. 2.4 Boxplots of total arsenic in market rice for country of production. *Solid black line* is mean, *red line* the median, the *outer edges* of the box are 25th and 75th percentile, *whiskers* 10th and 90th percentile and *dots* 5th and 95th percentile. N for each country is shown in parenthesis Australia (11), Bangladesh (99), China (98), Egypt (109), France (33), Ghana (35), India (70), Italy (27), Japan (26), Nepal (12), Pakistan (14), Spain (50), Thailand (49), USA (198) (Data are compiled primarily from Meharg et al. (2009) and Adamoko et al. (2011) with additional samples from the laboratories that produced these data also added in)

concentrations as Pakistani, the number of samples (N) for each country was only 4, compared to an N of 144 for Fig. 2.4 for Bangladesh. The median concentration in US rice reported by Zavala and Duxbury (2008) was ~170 ng g^{-1} (N = 112), which was lower than at 250 ng g^{-1} (N = 198) shown in Fig. 2.3.

When considering within country market basket rice it has to be considered if different regions of a country have different levels, as observed for Bangladesh (Duxbury et al. 2003; Meharg et al. 2009; Meharg and Rahman 2003; Zavala and Duxbury 2008), US (Williams et al. 2007a; Zavala and Duxbury 2008), Ghana (Adomako et al. 2011) and China (Williams et al. 2009; Zhu et al. 2008). It is important to establish countrywide frequency distribution of total arsenic that all major rice growing seasons are sampled and that weighting is given to relative production rates. For example, when Williams et al. (2006) sampled US market rice they did so at a frequency of the two major production areas with samples collected being proportionate to their relative production. Similarly, Williams et al. (2006) surveyed arsenic impacted and non-impacted regions of Bangladesh at approximate frequencies that equated to the importance of production from these two zones. This weighting of to relative production is crucial if regional, nation and global arsenic rice burdens are to be understood. Large within country variation in rice grain arsenic is observed for the US where the South-Central region has mean arsenic concentrations almost double that for California, 300 and 170 ng g^{-1}, respectively (Williams et al. 2007a).

Table 2.3 Descriptive statistics comparing wet (Aman) and dry (Boro) season market rice from 23 regions of Bangladesh which either have low (−As) or high (+As) arsenic in irrigation water analysed from data reported by Williams et al. (2006)

Irrigation water	Season	Mean grain As (ng g^{-1})	Std. Dev.	N
+As	Aman	182	75	11
	Boro	262	106	12
	Total	223	99	23
−As	Aman	136	41	8
	Boro	173	52	9
	Total	155	50	17
Total	Aman	162	66	19
	Boro	223	96	21
	Total	195	88	40

General linear modelling of the data showed that both As in groundwater (P=0.01) and season (P=0.023) were significant, but the interaction between the two factors was not (P=0.391)

Similar differences and values were observed by Zavala and Duxbury (2008) when comparing Texas (South Central region) and California, with Arkansas (also, South-Central) being intermediate.

Another example of within country geographic differences in total grain arsenic in market basket surveys is Bangladesh where irrigation with arsenic contaminated waters increases total grain arsenic, on average by 44% (Table 2.3). The Bangladeshi example is interesting as it also shows a strong and significant difference between wet season (Aman) and dry season (Boro) production (Williams et al. 2006). Arsenic contaminated groundwaters are used only during the dry season, but it may be expected that there is a residual carryover to the wet season and arsenic concentrations in rice grain are 27–44% higher in dry compared to wet season. However, regions of Bangladesh with low arsenic in groundwater also show this increase during the dry season and there is no significant interaction between season and arsenic in irrigation water (P=0.391) in the two-way analysis of variance of the data. This is best interpreted that dry season rice cultivation raises grain arsenic levels *per se* (Lu et al. 2009); the reasons are not entirely clear. In terms of human consumption of rice, this means that there are at least four defined Bangladesh rice types with respect to total grain arsenic, the lowest being wet season rice production in low groundwater arsenic areas with a mean of 136 ng g^{-1}, and the highest being dry season high groundwater arsenic region at 262 ng g^{-1}, representing about twofold difference in rice arsenic exposure. From Fig. 2.2 it can be seen that grain inorganic arsenic concentration for Bangladeshi rice is linear with total arsenic over this range, so this doubling of grain total arsenic equates directly to a doubling of grain inorganic arsenic intake.

Though less systematically observed than for the US and Bangladesh examples, it is clear that market rice is also elevated in arsenic in base and precious metal mining regions of SE Asia (Williams et al. 2009; Zhu et al. 2008) and Africa (Adomako et al. 2011).

When Zavala and Duxbury (2008) calculated a "global average" for total grain arsenic concentration, they simply averaged available datasets, not accounting for the fact that a large and disproportionate sample, relative to global production, were from the US. In contrast, Meharg et al. (2009) calculated global distributions of arsenic in rice by weighting each country's total arsenic distribution by that country's relative contribution to global rice production.

Market rice, by essence, averages, at least over an entire production season if not individual batches, grain arsenic emanating from a particular agronomic zone, the size of which is determined by the relationship of mills to farmers' fields. As complex distribution chains amalgamate outputs from individual mills and transport that rice to major population centres, the homogeneity of grain should increase with increasing size of the population being supplied with that rice. Thus, local subsistence farmers who are consuming their own grain, or at least have that grain homogenised with others at local mills, should observe most variation. Field studies should show wider variation in total arsenic concentration than market basket studies, and this is indeed the case (Lu et al. 2009; Williams et al. 2009; Zhu et al. 2008). When considering arsenic exposure to the total population, market surveys are appropriate; but when thinking about risks to subpopulations living in the most arsenic contaminated areas, a more localized consideration of arsenic in rice is needed. This is explored further in Chap. 4 where the sources of arsenic to paddy fields are considered in detail.

2.5 Rice Products

Rice products, due to rice having good textural properties and being bland tasting, are prevalent in western diets and may be the primary route of rice ingestion. They include rice based breakfast products (crisped and puffed), cereal bars and a wide range of processed food. Rice is gluten free which makes it a staple for those with gluten allergies and, along with its blandness and textural properties, this also makes it suitable for weaning infants where it is advised to avoid foods that may cause allergies in later life. Indeed, rice porridge is the first food encountered by most humans (Meharg et al. 2008a). Later stage weaning products often have a rice porridge base to them as well. Certain subpopulations have a high reliance on rice products, besides cooked rice grain. These include, again, those with gluten intolerance mentioned above where rice based breads, cakes and snacks predominate in catering for gluten free dietary requirements. Other groups of food intolerance with high rates of rice consumption are lactose intolerance sufferers who use vegetable milk substitutes (Meharg et al. 2008b). Soya milk used to dominate this market but rice milk has, to many, better taste properties to some (it is blander) and does not have elevated phyto-hormones that typify soya milk and which raise health concerns amongst some. These health concerns are for cancer patient sufferers, and breast cancer patients in particular are advised to use rice milk as part of their diet to avoid hormones associated with animal and soya milk. Vegan and macrobiotic diets have a high reliance on rice products such as rice milk, rice based confectionary (milk substitute and rice molasses), condiments (miso, rice wine vinegar etc.), cooking oil (bran oil) and

Table 2.4 Total and inorganic arsenic concentrations in rice products purchased commercially

Product	N	Total arsenic (mg kg^{-1})	Inorganic arsenic (%)
Amazake	3	0.14	74
Baby rice (pure porridge)	17	0.11	53
Bran	8	0.89	74
Bran oil (for frying)	3	0.03	–
Candy	5	0.14	–
Crackers (biscuits)	11	0.28	75
Crisped (breakfast cereal)	3	0.21	86
Koji	3	0.12	63
Malt	3	0.21	
Milk	9	0.02	67
Mirin	2	0.01	
Miso	3	0.12	71
Molasses	2	0.21	81
Noodles	6	0.12	80
Puffed (breakfast cereal)	2	0.24	77
Vinegar	4	0.05	–
Wine	4	0.02	–

Data are compiled from Sun et al. (2008, 2009), Signes-Pastor et al. (2009), Meharg (2007) and Meharg et al. (2008a, b)
A "–" means not analysed

sugar (molasses) (Signes-Pastor et al. 2009; Sun et al. 2009). Much of modern vegan, macrobiotic and, to a lesser extent, vegetarian dietary practices are derived from traditional Japanese diets. Recently, there has been commercial interest in rice bran. Rice bran is used in traditional cooking such as some Japanese pickling recipes, but it has been marketed as both a health food and as a nutritional supplement for the under-nourished (Sun et al. 2008). So rice based food products are diverse in nature and in this section their total and inorganic arsenic concentrations will be considered.

The arsenic concentration of liquid rice products (bran oil, mirin, milk, vinegar and wine) are an order of magnitude lower than the solid products (Table 2.4). With the exception of rice milk, and perhaps rice wine, the other liquid products are consumed in relatively small volumes. Rice milk consumption, particularly in toddlers, can be high on a body mass basis (Meharg et al. 2009), and although arsenic levels, most of which is inorganic, is lower than solid products, the concentrations actually exceed WHO, US-EPA and EU drinking water standards (10 μg L^{-1}). As young children get much of their fluid intake from milk, or milk substitutes if they are lactose intolerant, these levels of arsenic in rice milk have raised concern. Indeed, the UK's Food Standard Authority has issued advice (FSA 2009) that children under 4 years of age should not take rice milk. Amazake (liquid fermented rice of gruel consistency) could be considered a drink as it is fluid, but is more akin to yogurt or cream in the quantities and occasions for consumption.

Most solid rice products have average arsenic concentrations of 0.1–0.2 mg kg^{-1}, typical of raw rice (see Table 2.4). Although a particular individual may not eat a

Table 2.5 Total arsenic concentrations (mg kg^{-1}) in EU cereal grains from the EFSA (2009) review

Food type	N		Median	Mean
Cereal excluding rice	2,215	LB	0.000	0.015
		UB	0.026	0.040
Rice grain	1,122	LB	0.110	0.136
		UB	0.110	0.142
Cereal products excluding rice	1,004	LB	0.000	0.011
		UB	0.012	0.030
Rice products	314	LB	0.100	0.142
		UB	0.100	0.166
Bran and germ	13	LB	1.630	2.134
		UB	1.630	2.134
Cereal grains and cereal products	5,047	LB	0.000	0.054
		UB	0.0400	0.072

The Lower Bound (LB) is obtained by assigning a value of zero (minimum possible value) to all the samples reported as <LOD or <LOQ. The Upper Bound (UB) is obtained by assigning the value of LOD to values reported as <LOD and LOQ to values reported as <LOQ (maximum possible value), depending on whether LOD or LOQ is reported by the laboratory

lot of raw ice, rice consumption could still be high through widely available and popular products such as breakfast cereals, crackers and cereal bars with high rice content. Rice flour, molasses and malt are widely used in processed foods, particularly for the health food market.

Rice bran has the highest levels of total and inorganic arsenic (Sun et al. 2008). The risks from this product will be considered in Chap. 3. Here it is suffice to say that rice bran products are used as nutritional supplements as well as finding itself as a component of health foods, such as rice crackers, which may explain the high arsenic content of crackers.

When calculating dietary exposure from rice, these processed products must also be considered. The arsenic concentration of these products will vary based on the raw rice used to construct them, and the raw rice varies in its arsenic concentration regionally (Table 2.2). For foods of particular concern, such as baby ice and rice milk, grain to produce these products could be sourced from low arsenic regions.

The European Food Safety Authority has compiled arsenic analysis in foods throughout the EU (EFSA 2009) and Table 2.5 reports the values for cereal grains, based on >5,000 individual analysis. It can be seen that rice and rice products dominate arsenic levels in cereals, by an order of magnitude.

References

Abedin J, Cresser M, Meharg AA, Feldmann J, Cotter-Howells J (2002) Arsenic accumulation and metabolism in rice (*Oryza sativa* L.). Environ Sci Technol 36:962–968

Adomako EE, Williams PN, Deacon C, Meharg AA (2011) Inorganic arsenic and trace element intake from locally produced versus imported rice. Environ Pollut 159:2435–2442

Arao T, Maejima Y, Koji B (2009) Uptake of aromatic arsenicals from soil contaminated with diphenylarsinic acid by rice. Environ Sci Technol 43:1097–1101
Baba K, Arao T, Maejima Y, Watanabe E, Eun H, Ishizaka M (2008) Arsenic speciation in rice and soil containing related compounds of chemical warfare agents. Anal Chem 80:5768–5775
Carey AM, Scheckel KG, Lombi E, Newville M, Choi Y, Norton GJ, Charnock JM, Feldmann J, Price AH, Meharg AA (2010) Grain unloading of arsenic species in rice (*Oryza sativa* L.). Plant Physiol 150:309–319
de la Calle MB, Emteborg H, Linsinger TPJ, Montoro R, Sloth JJ, Rubio R, Bacter MJ, Feldmann J, Vermaercke P, Raber G (2011) Does the determination of inorganic arsenic in rice depend on method? Treds Analy Chem 30:641–651
Duxbury JM, Mayer AB, Lauren JG, Hassan N (2003) Food chain aspects of arsenic contamination in Bangladesh: effects on quality and productivity of rice. J Environ Sci Health A Tox Hazard Subst Environ Eng 38:61–69
EFSA (2009) Panel on Contaminants in the Food Chain (CONTAM)/Scientific opinion on arsenic in food. EFSA J 7:1351
Feldmann J, Salaun P, Lombi E (2009) Critical review perspective: elemental speciation analysis methods in environmental chemistry – moving towards methodology. Environ Chem 6:275–289
Freitas MC, Pacheco AMG, Bacchi MA, Dionisio I, Landsberger S, Fernandes EAN (2008) Compton suppression instrumental neutron activation analysis performance in determining trace- and minor-element contents in foodstuff. J Radioanal Nucl Chem 276:149–156
FSA (2009) http://www.food.gov.uk/multimedia/pdfs/fsis0209arsenicinrice.pdf. Last Accessed on 02/01/2012
Hansen HR, Raab A, Price AH, Duan G, Zhu Y, Norton GJ, Feldmann J, Meharg AA (2011) Identification of tetramethylarsonium in rice grains with elevated arsenic content. J Environ Monit 13:32–34
Heitkemper DT, Vela NP, Stewart KR, Westphal CS (2001) Determination of total and speciated arsenic in rice by ion chromatography and inductively coupled plasma mass spectrometry. J Anal Atom Spectrom 16:299–306
Jorhem L, Astrand C, Sundstrom B, Baxter M, Stokes P, Lewis J, Grawe KP (2008) Elements in rice from Swedish market: 1. Cadmium, lead and arsenic (total and inorganic). Food Addit Contam 25:284–292
Juhasz AL, Smith E, Weber J, Rees M, Rofe A, Kuchel T, Sansom L, Naidu R (2006) In vivo assessment of arsenic bioavailability in rice and its significance for human health risk assessment. Environ Health Perspect 114:1826–1831
Lamont WH (2003) Concentration of inorganic arsenic in samples of white rice from the United States. J Food Compos Anal 16(6):687–695. doi:10.1016/S0889-1575(03)00097-8
Li RY, Ago Y, Liu WJ, Mitani N, Feldmann J, McGrath SP, Ma JF, Zhao FJ (2009) The rice aquaporin Lsi1 mediates uptake of methylated arsenic species. Plant Physiol 150:2071–2080
Lin HT, Wong SS, Li GC (2004) Heavy metal content of rice and shellfish in Taiwan. J Food Drug Anal 12:167–174
Liu CW, Lin CC, Jang CS, Sheu GR, Tsui L (2009) Arsenic accumulation by rice grown in soil treated with roxarsone. J Plant Nutr Soil Sci 172:550–556
Lombi E, Susini J (2009) Synchrotron-based techniques for plant and soil science: opportunities, challenges and future perspectives. Plant Soil 320:1–35
Lombi E, Scheckel KG, Pallon J, Carey AM, Zhu YG, Meharg AA (2009) Speciation and distribution of arsenic and localization of nutrients in rice grains. New Phytol 184:193–201
Lu Y, Adomako EE, Solaiman ARM, Islam RM, Deacon C, Williams PN, Rahman GKMM, Meharg AA (2009) Baseline soil variation is a major factor in arsenic accumulation in Bengal Delta paddy rice. Environ Sci Technol 43:1724–1729
Meharg AA, Hartley-Whitaker J (2002) Arsenic uptake and metabolism in arsenic resistant and non-resistant plant species. New Phytol 154:29–43

Meharg AA, Raab A (2010) Getting to the bottom of arsenic standards and guidelines. Environ Sci Technol 44:4395–4399

Meharg AA, Rahman Md M (2003) Arsenic contamination of Bangladesh paddy field soils: implications for rice contribution to arsenic consumption. Environ Sci Technol 37:229–234

Meharg AA, Sun G, Williams PN, Adamako E, DeaconC ZhuYG, Feldmann J, Raab A (2008a) Inorganic arsenic levels in baby rice are of concern. Environ Pollut 152:746–749

Meharg AA, Deacon C, Campbell RCJ, Carey A-M, Williams PN, Feldmann J, Raab A (2008b) Inorganic arsenic levels in rice milk exceed EU and US drinking water standards. J Environ Monit 10:428–431

Meharg AA, Williams PN, Schekel K, Lombi E, Feldmann J, Raab A, Zhu YG, Gault A, Islam R (2008c) Speciation of arsenic differs between white and brown rice grain. Environ Sci Technol 42:1051–1057

Meharg AA, Williams PN, Adamako E, Lawgali YY, Deacon C, Villada A, Cambell RCJ, Sun GX, Zhu YG, Feldmann J, Raab A, Zhao FJ, Islam R, Hossain S, Yanai J (2009) Geographical variation in total and inorganic arsenic content of polished (white) rice. Environ Sci Technol 43:1612–1617

Munoz O, Diaz OP, Leyton I, Nunez N, Devesa V, Suner MA, Velez D (2002) Vegetables collected in the cultivated Andean area of northern Chile: total and inorganic arsenic contents in raw vegetables. J Agric Food Chem 50:642–647

National Research Council (2001) Arsenic in drinking water – 2001 update. National Academy Press, Washington, DC

Norton GJ, Islam MR, Deacon CM, Zhao FJ, Stroud JL, McGrath SP, Islam S, Jahiruddin M, Feldmann J, Price AH, Meharg AA (2009) Identification of low inorganic and total grain arsenic rice cultivars from Bangladesh. Environ Sci Technol 43:6024–6030

Raab A, Feldmann J, Meharg AA (2004) The nature of arsenic – phytochelatins complexes in *Holcus lanatus* and *Pteris cretica*. Plant Physiol 134:1113–1122

Raab A, Baskaran C, Feldmann J, Meharg AA (2008) Cooking rice in a high water to rice ratio reduces inorganic arsenic content. J Environ Monit 11:41–44

Schoof RA, Yost LJ, Eickhoff J, Crecelius EA, Cragin DW et al (1999) A market basket survey of inorganic arsenic in food. Food Chem Toxicol 37:839–846

Signes-Pastor AJ, Deacon C, Harris P, Carbonell-Barrachina AA, Meharg AA (2009) Arsenic speciation in Japanese rice drinks and condiments. J Environ Monit 11:1930–1934

Sun GX, Williams PN, Carey AM, Zhu YG, Deacon C et al (2008) Inorganic arsenic in rice bran and its products are an order of magnitude higher than in bulk grain. Environ Sci Technol 42:7542–7546

Sun G-X, Zhu YG, Williams PN, Deacon C, Meharg AA (2009) Survey of arsenic and its speciation in cereal rice products. Environ Int 35:473–475

Torres-Escribano S, Leal M, Velez D, Montoro R (2008) Total and inorganic arsenic concentrations in rice sold in Spain, effect of cooking, and risk assessments. Environ Sci Technol 42:3867–3872

Williams PN, Price AH, Raab A, Hossain SA, Feldmann J, Meharg AA (2005) Variation in arsenic speciation and concentration in paddy rice related to dietary exposure. Environ Sci Technol 39:5531–5540

Williams PN, Islam MR, Adomako EE, Raab A, Hossain SA et al (2006) Increase in rice grain arsenic for regions of Bangladesh irrigating paddies with elevated arsenic in groundwaters. Environ Sci Technol 40:4903–4908

Williams PN, Raab A, Feldmann J, Meharg AA (2007a) Market basket survey shows elevated levels of as in South Central US processed rice compared to California: consequences for human dietary exposure. Environ Sci Technol 41:2178–2183

Williams PN, Villada A, Deacon C, Raab A, Figuerola J et al (2007b) Greatly enhanced arsenic shoot assimilation in rice leads to elevated grain levels compared to wheat and barley. Environ Sci Technol 41:6854–6859

Williams PN, Lei M, Sun G-X, Huang Q, Lu Y, Deacon C, Meharg AA, Zhu Y-G (2009) Occurrence and partitioning of cadmium, arsenic and lead in mine impacted paddy rice: Hunan, China. Environ Sci Technol 43:637–642

World Health Organization (2004) IARC, Working Group on some drinking water disinfectants and contaminants, including arsenic, vol 84. Lyon. Monograph 1

Xu XY, McGrath SP, Meharg A, Zhao FJ (2008) Growing rice aerobically markedly decreases arsenic accumulation. Environ Sci Technol 42:5574–5579

Zavala YJ, Duxbury JM (2008) Arsenic in rice: I. Estimating normal levels of total arsenic in rice grain. Environ Sci Technol 42:3856–3860

Zavala YJ, Gerads R, Gürleyük H, Duxbury JM (2008) Arsenic in rice: II. Arsenic speciation in USA grain and implications for human health. Environ Sci Technol 42:3861–3866

Zhao FJ, Ma JF, Meharg AA, McGrath SP (2009) Arsenic uptake and metabolism in plants. New Phytol 181:777–794

Zhu Y-G, Sun G-X, Lei M, Teng M, Liu Y-X, Chen N-C, Hong W-L, Carey AM, Meharg AA, Williams PN (2008) High percentage inorganic arsenic content of mining impacted and nonimpacted Chinese rice. Environ Sci Technol 42:5008–5013

Chapter 3
Risk from Arsenic in Rice Grain

3.1 Rice Consumption

Rice consumption rates, on average, vary by nearly 4 orders of magnitude (0.9–650 g/person/d) for different countries, with Myanmar, Laos, Vietnam, Cambodia and Bangladesh, clustered around SE Asia, having the highest consumption rates (Fig. 3.1). At the other extreme, very low consumption rates of rice are found in some African and European countries. Forty-six countries consumed more than 100 g of rice per person per day, and these included African, South American and Asian countries. These data were compiled from UN FAO data (FAO 2004) on rice production, imports and exports per annum, divided by per capita and, as a result, will smooth data sub-structure for each country, as rice consumption will vary greatly between individuals. Also, it is not the rice consumption alone that drives arsenic exposure from rice, but consumption per unit body mass, adding a further variable to consider. While simple averages can be used to come up with rough estimates of exposure, i.e. average inorganic arsenic concentration of rice, average intake of that rice and average body mass, as conducted by Meharg et al. (2009) and Zavala et al. (2008), the most robust approach is to gather detailed information for populations of interest, divided into meaningful subclasses (male/female, age substructure etc.) and then use probabilistic approaches to calculate not only average intakes, but also variance of these intakes and percentiles of interest, such as conducted by Mondal and Polya (2008) for West Bengal, India populations.

For countries that are quite uniform in their rice eating habit, estimating dietary intakes of arsenic is relatively straightforward. Examples of this are Bengal Delta populations in West Bengal, India and Bangladesh where rice-eating culture predominates. However, India illustrates the point where country averages can be misleading as, although the rice eating culture and consumption rates in West Bengal are approximately the same as Bangladesh, India, on average, consumes half the amount of rice per capita of Bangladesh. This lower rice consumption for India is due to it having large wheat eating populations to the northwest.

A.A. Meharg and F.J. Zhao, *Arsenic & Rice*, DOI 10.1007/978-94-007-2947-6_3,

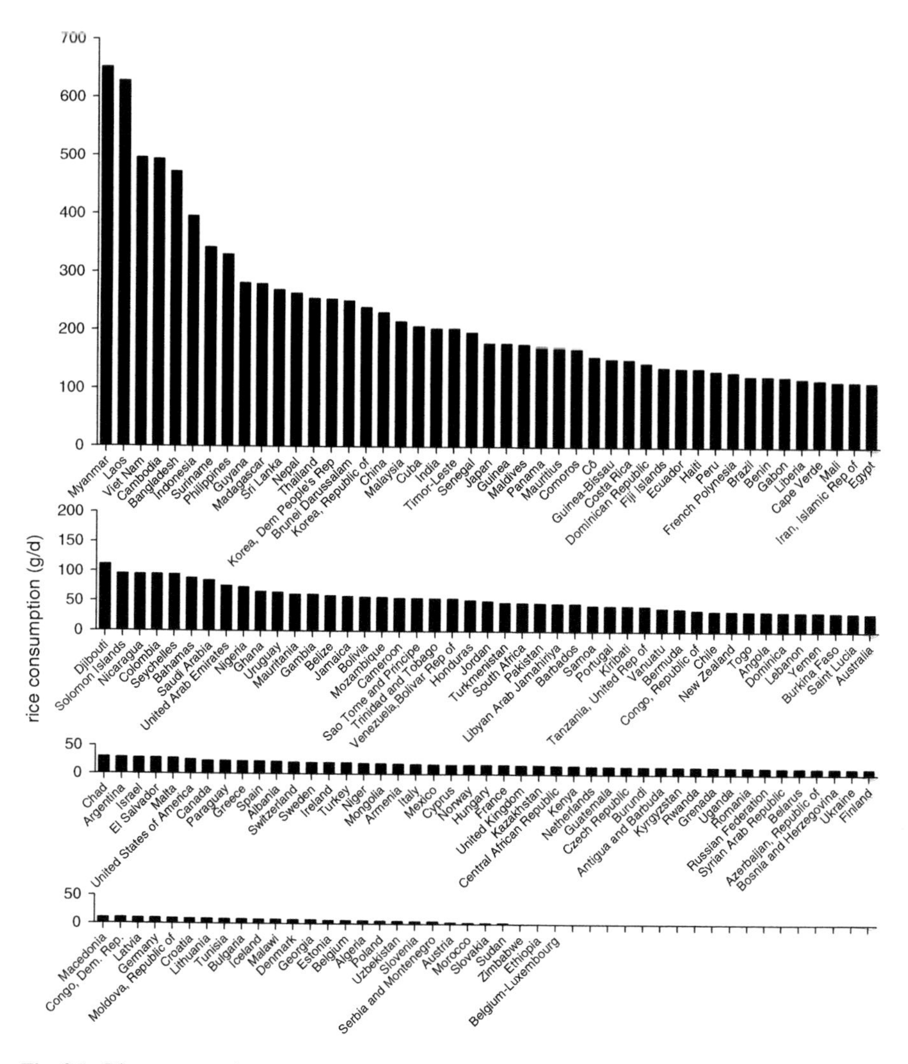

Fig. 3.1 Rice consumption rates by country calculated from FAO (2004)

Immigration further complicates matters as minority rice eating cultures in non-traditional rice eating areas, such as Western Europe or North America, can have high rates of rice consumption. As risk assessment should be targeted at those at the highest exposure, subpopulations eating high rice diets need to be considered as special cases. Detailed national data are analysed for the UK to illustrate how ethnicity and age affect rice consumption rates. The data are obtained from Department of Environment, Food and Rural Affair's Expenditure and Food Survey (EFS), which reports household purchase of rice on a per capita basis, and from the National Diet and Nutrition Survey (NDNS), reporting daily food consumption per person (National Diet and Nutritional Survey 1992, 1997, 2000, Expenditure and Food Survey).

Table 3.1 Daily adult rice purchase (g/d) from the Expenditure and Food Survey database broken down into detailed ethnic grouping (Meharg 2007)

Description	Total rice purchased (g/d)
White-British	8
Other white background	10
Other mixed background	15
Asian-Indian	28
Asian-Pakistani	30
Black African	32
Chinese	35
Black Caribbean	36
Other black background	48
Other Asian background	61
Other ethnic background	118
Asian-Bangladeshi	251

Data from DEFRA Expenditure and Food Survey. Averages for the 3 years ended 31st March 2005, Updated on: 25/05/2006. Supplied by the Food Statistics Branch

The EFS data enables rice consumption based on ethnicity to be considered. In Table 3.1, data are expressed as daily purchase (assumed to equal consumption) and highlight that different ethnic classes consume considerably differing quantities of rice. The largest purchasing group was Asian-Bangladeshi who purchased total rice amounting to ~250 g/d, >30 times more than the average White-British person. The next largest rice consuming sub-population is "other ethnic background" at ~120 g/d, and includes people of Middle Eastern origin, South Americans and Pacific Islanders amongst others, many of whom come from countries where rice is a staple (Fig. 3.1). White-British had the lowest rates of rice consumption. Bangladeshi's make up 0.5% of the UK population, and the "other" category accounts for 0.4%. High rice (>100 g/d) consuming sub-populations make up 0.9% of the UK populace.

The differences in rice purchase rates between different components of the UK's Asian population was also found by Wharton et al. (1984) who investigated food intake patterns in pregnant Moslems, Sikhs and Hindus in a maternity hospital. They report that rice was the most common food eaten at 10% or more of meal times. For Pakistanis and Sikhs rice did not rise above 10% and for Hindus this figure was 25% for lunch and 38% for the evening meal. For Bangladeshis, 100% consumed rice at lunch and 69% in the evening. Emigration appears to have decreased traditional dietary intakes as rice consumption in Bangladesh varies from 400 to 650 g/d (Rahman et al. 2008), while in SE Asian 180–300 g/d is typical (Lin et al. 2004; Liu et al. 2005).

The data held in the NDNS records do not consider ethnicity, but are broken into percentiles with age structure within the dataset: adult, young person (4–18 y) and toddler (1.5–4.5 y) (Table 3.2). The 99th percentile for 18 year olds (134 g/d) is 53% lower than average Bangladeshi rice purchase (251 g/d) (Table 3.1). As Bangladeshis represent 0.5% of the UK populace, it makes sense that their rice intake is higher than the UK 99th percentile, thus the NDNS and EFS databases confirm each other. Body weight data are from the survey itself, based on respondent's information. Expressing

Table 3.2 Total daily raw rice consumption rates[a] from National Diet and Nutrition Survey (NDNS) (Meharg 2007)

Percentile	Consumption per person (g/d)	Consumption per kg body weight (g/kg/d)
Mean		
>18 years	23	0.3
4–18 years	23	0.6
1.5–4.5 years	14	1.0
95th percentile		
>18 years	76	1.0
4–18 years	74	2.2
1.5–4.5 years	42	3.0
99th percentile		
>18 years	134	1.8
4–18 years	170	3.6
1.5–4.5 years	63	4.8

[a]Data in the Table are the sum of boiled rice and raw daily rice consumption converted into raw rice, using a 3:1 boiled:raw rice conversion factor. Body masses of 65 kg for adult, 33.8 kg for young person and 12.5 kg for toddler were used in the consumption on a body weight basis calculation. Body masses were directly recorded from respondents

Table 3.3 Rice consumption rates (dry weight) in the US from the US CSFII database considered by ethnic origin, from Batres-Marquez and Jensen (2005)

	Percentage of overall average (%)	Percentage of rice consumer average (%)
White, non-Hispanic	64	80
Black, non-Hispanic	132	103
Mexican-American	110	85
Other Hispanic	170	104
Other	503	207
Born outside US	338	147

rice consumption rates on a body weight basis highlights that children, particularly toddlers, have higher exposures to arsenic from rice than adults (Table 3.2).

The US Department of Agriculture Continuing Survey of Food Intake by Individuals (CSFII) database reports similar information to UK databases (Tsuji et al. 2007), though the data are graphically reported and on a wet weight (i.e. boiled rice), making direct comparisons difficult. A more detailed comparison is possible with data presented by Batres-Marquez and Jensen (2005) who report US rice consumption rates on a dry weight basis using the CSFII survey (Table 3.3). White, non-Hispanic consumed the least amount of rice. The group "others" consisting of Asians, Pacific Islanders and Native Americans ate more than 115 g dry weight (d. wt.) of rice per day compared to the average US intake, which included non-rice consumers, of 11 g/d. d. wt. For those who did consume rice the average consumption rate was 61 g/d d. wt. They found that being born outside the US, i.e. first generation immigrants, had higher rice consumption rates than average.

Another finding was that those on lower incomes had higher rates of rice consumption. When performing risk assessments related to inorganic arsenic intake from rice (see Sect. 3.5), it is clear that such assessments need to have clear demographical breakdown of the studied populations.

As mentioned in Sect. 2.5, other sub-populations, based on health and/or lifestyles, may consume high amounts of rice and rice products. These "health" sub-populations are probably dominated by gluten intolerances and Celiac disease sufferers, who tend to substitute wheat and other gluten containing substances with rice as the most palatable alternative source of carbohydrate (Thompson 2001). Celiac disease is prevalent in Northern European, and affects 1 in 133 of the US population. Rice products such as rice biscuits, rice wafers, rice pasta, crisped ice cereals etc., are specifically marketed to target those with gluten intolerances. Those who actively choose some specific diet plans such as vegans and macrobiotic, will have a higher exposure to rice. For example, macrobiotic diets recommend rates of rice consumption normally comprising two meals a day, with a high reliance on fermented rice products and the use of vegetable milk.

3.2 Studies on the Total Intake of Inorganic Arsenic

Schoof et al. (1999) provided the first substantial indication that dietary exposure to inorganic arsenic could be problematic, concluding that diets that rely heavily on rice may contain the most inorganic arsenic. In their food basket survey rice greatly exceed all other foods tested, the next lowest foods (flour [unspecified], watermelon and grape juice) were sevenfold or more lower in inorganic arsenic. Fish products such as tuna and shrimp had very high total arsenic levels, but contained at least 30-fold lower inorganic arsenic than rice. However, Schoof et al. (1999) did not assess intake from particular food types.

The European Food Safety Authority (EFSA) presented similar findings that comprehensively reviewed inorganic arsenic in European diets (EFSA 2009). The review found that at median consumption rates rice, the dominant contribution to the "grain and cereal" (Table 2.5), accounted for over 50% of dietary exposure to inorganic arsenic (Fig. 3.2). This is in sharp contrast to water, which was assumed to be the dominant exposure route until problems regarding inorganic arsenic in rice were reported, accounted for less than 1.5% of intake. Unfortunately, the EFSA review does not report the food intakes for which the data presented in Fig. 3.2 were collected.

Similar approaches to the EFSA have been conducted for USA, based on the CSFII database, extrapolated with limited measurements of arsenic in foods, or smaller dietary intake studies for specific regions, and have their limitations in that that they did not have as full information regarding arsenic speciation in foods. The CSFII was used to estimate dietary intakes of inorganic arsenic in adults, which ranged from 1.8 to 11.4 μg/d for males and 1.3–9.4 μg/d for females at the 10th and 90th percentile, respectively (Meacher et al. 2002). Mean intake was 6.3 ± 10.6 and 5.2 ± 9.3 μg/d for males and females respectively. The 95th percentile for both men and women is about 15.9 and 13.2 μg/d for men and women respectively. In three out of the four US

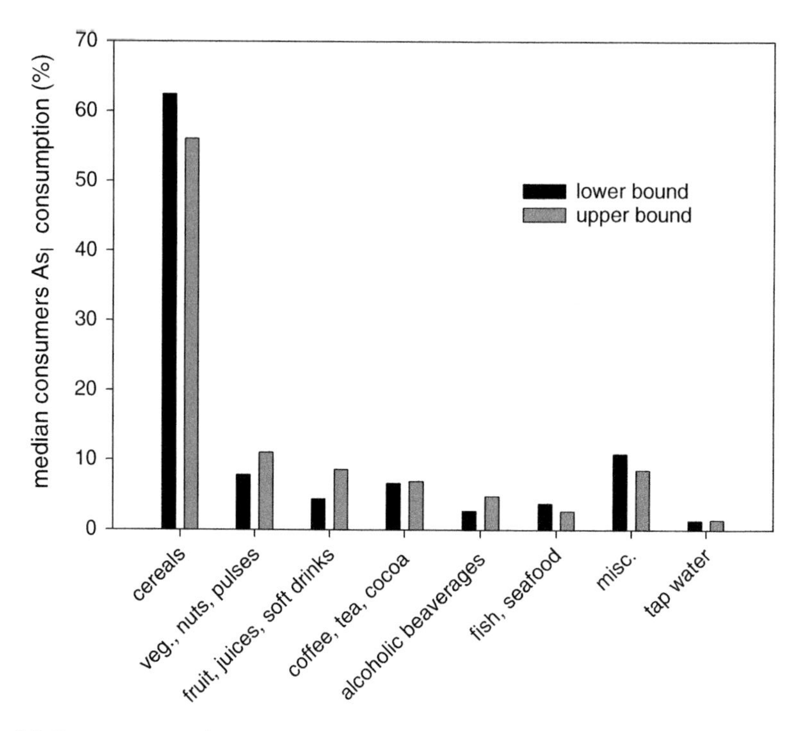

Fig. 3.2 Percentage contribution of different food sources to median dietary intake of inorganic arsenic (As$_i$) in the EU from EFSA commissioned report (EFSA 2009)

regions with low arsenic in drinking water, food accounted for over 50% of dietary intakes of inorganic arsenic. No presentation of where the arsenic in food was coming from was detailed. A similar study was conducted by Tao and Bolger (1999). This study was somewhat limited in that the total arsenic levels in rice they report, 0.03–0.11 mg/kg, are very low compared to other studies (see Sect. 2.4). Also, they assumed that 100% of the total arsenic in rice was inorganic, except for seafoods where 10% inorganic arsenic was assumed. US rice has approximately 50% inorganic arsenic in the total (see Sect. 2.3), whereas for seafood the percentage is much lower than 10% (Schoof et al. 1999). Tsuji et al. (2007) modelled exposure to the US populace again using the CSFII data and based inorganic arsenic concentration in foods on Yost et al. (2004), and considered the 40 food types that made up to 90% of US exposure according to Schoof et al. (1999). The Foods and Residue Evaluation (FARE) model was used in modelling. At the 95th percentile inorganic arsenic intake from rice was 6.1 μg/d for adults and 3.3 μg/d for children of 1–6 years, by far the highest food source, and similar to intake from water when exposure was truncated to below 0.01 mg/L As for water. The mean values for children and adults from rice were 1 and 1.67 μg/d, respectively. Meliker et al. (2006) conducted a study on 440 adults from the state of Michigan, USA, recording what they ate and drunk, and then used literature estimates of the inorganic arsenic concentration of these foods and drink to estimate dietary exposure. However, the study was not structured with respect to demography as more than 80% of participants were over 60 years old, 87% were male and over 90% were white, so the results must be interpreted with this in mind. In particular, only 2% of the study group was African America/Black while the population of this group in Michigan counties studied was 10%. Similarly the study group contained only 2% Asian/Asian American with the actual population being 6%. In the regions of Michigan with low arsenic in drinking water, 57% of inorganic arsenic ingestion was from food "with almost all the intake from rice". The authors found 8.1, 32.4, 32.4 and 793.8 g/day rice consumption for the 10th, 50th, 90th and maximum intake, respectively. The percentiles for inorganic arsenic concentration of rice used in this modelling were 0.05, 0.11, 0.18 and 0.27 mg/kg, respectively, with the data obtained from Lamont (2003).

Actual measures of dietary intake for small cohorts have been ascertained, primarily for the Bengal Delta (Kile et al. 2007; Mondal and Polya 2008; Ohno et al. 2007), and this is the most accurate way of assessing arsenic intake in a particular populations, although it only gives intakes for the population under study. Intake rates for such direct measurement studies are in general agreement with studies looking at country averages. The scenarios are more complex in heterogeneous cultures, such as Europe and the US, where there are a lot more food choices, which also vary greatly between subpopulations. In essence, where a country has rice as the dominant staple, average population data such as that presented in Fig. 3.1 will give a good guide to the risks posed by the rice eating populace. For heterogeneous cultures with respect to rice consumption, detailed breakdown in sub-population eating habits is required to identify those at risk from high ingestion rates of inorganic arsenic from rice.

Mondal and Polya (2008) probabilistically modelled the contribution of rice and drinking water exposures in arsenic affected region of West Bengal, India and found that rice contributed 44% to inorganic arsenic exposure with drinking water 48%. A further 8% was contributed by rice cooking water (see Sect. 3.4 for a discussion of rice cooking). The arsenic level in drinking and cooking water was 17 μg/L, which is low for arsenic affected regions of the Bengal Delta (Smedley and Kinniburgh 2002), even though the authors were working in a designated arsenic affected region. The lower than expected values of arsenic in water probably were due to remedial technologies to reduce arsenic in the drinking water supply. Similarly, Ohno et al. (2007) studied an arsenic affected region of Bangladesh and found water arsenic levels lower than expected with a median of 20 μg/L. Here rice accounted for 56% of dietary arsenic intake (note: samples not speciated for arsenic). In another Bangladeshi study, this time in a low arsenic drinking water region (median arsenic 1.6 μg/L), 92% of arsenic came from food, with rice dominating food intake, and 82% of the arsenic present in food being inorganic arsenic species. In another study, where water arsenic ranged from 200 to 500 μg/L, cooked rice contributed 14% to dietary intake (Smith et al. 2006). It is clear for the Bengal Delta population that rice is a dominant source of inorganic arsenic exposure where water arsenic levels are low, and that the intake levels are elevated enough to constitute notable lifetime excess cancer risks.

Another West Bengal study looked at contribution of food composites for adult meals consisting of 750 g of rice per day across low (3 μg/L), intermediate (30 μg/L) and high (120 μg/L) arsenic in drinking water supply (Uchino et al. 2006). They found that intake of total arsenic was 8, 60 and 330 μg/d from water and 130, 170 and 200 μg/g from food, respectively for the three drinking water classes. Another study for West Bengal estimated 170 μg/d inorganic arsenic intake from rice (Signes-Pastor et al. 2008). Again, that where drinking water is relatively low, rice greatly dominates arsenic intake, and even when drinking water is high, rice is still a major contributor to dietary intake. To place the Bengali data in context, Ruangwises and Saipan (2010) conducted a similar study for Thai arsenic consumption and found inorganic arsenic intake from food of 80 μg/d, with rice consumption rates just over 200 g/d (Fig. 3.1). Dietary intakes of inorganic arsenic from rice were estimated at 58 μg/d consuming 460 g/d rice for a Vietnamese Red River Delta population (Agusa et al. 2009).

It can be seen that rice is the main dietary source of inorganic arsenic for both Thais and Bangladeshi cohorts on moderate to low arsenic in drinking water. The difference is just that Thais eat two to three times less rice than Bangladeshis, again highlighting rice as the dominant source of inorganic arsenic into the diet.

3.3 Considering Rice as an Arsenic Source in Epidemiological Studies

Inorganic arsenic, as indeed DMA, has been somewhat ignored when considering food and drinking water exposures. Argos et al. (2010) completed a detailed monitoring of inorganic arsenic ingestion from drinking water in Bangladeshis over 9 years, and used this information along with health outcomes, to predict disease dose responses. Because they ignored rice as a dietary source of inorganic arsenic, the assumptions they made regarding inorganic arsenic intakes and disease outcomes are inappropriate. Particularly pertinent was the fact that the study was looking at the effects of switching from high drinking water arsenic to low, where a range of studies have shown for the Bengal delta that this switch would make rice as the dominant source of inorganic arsenic exposure (Kile et al. 2007; Meharg et al. 2009; Mondal and Polya 2008; Ohno et al. 2007).

Another example of where not considering arsenic intakes from rice can lead to misinterpreting results is to consider urinary arsenic speciation without taking rice as a dietary source into account. Navas-Acien et al. (2011) undertook the considerable task of speciating >4,000 US urine samples and then related this information to seafood intakes and tried to interpret their findings with respect to inorganic arsenic and DMA concentration in urine to metabolism of seafood. However, they had ignored rice consumption, where rice speciation is dominated by inorganic arsenic (which is readily metabolised by humans to DMA) and DMA (Williams et al. 2005). In a smaller study where ethnicity and dietary consumption patterns were studied between UK Bangladeshis and Caucasians, differences in rice consumption lead to differences in urinary DMA and inorganic arsenic (Casini et al. 2011). This UK study also measured arsenobetaine and found it to be much higher in Caucasians compared to Bangladeshis, as the latter prefer freshwater fish, which is low in arsenobetaine. The US study (Navas-Acien et al. 2011) showed a positive correlation between arsenobetaine and DMA. Many people eat rice and seafood together. Seafood and rice sources of arsenic need to be untangled before any pronouncement on what are the sources of arsenic that end up in humane urine.

Similarly, Mohri et al. (1990) ignored rice as a dietary source of arsenic species in Japanese diet, even though Japanese consume relatively large quantities of rice (Fig. 3.1). They came up with 10 μg/d intake of inorganic arsenic, which is probably a considerable underestimate. A Thai study looked at the content of inorganic arsenic in a composite diet, and while the composition of that composite was given, it can be assumed that rice was prevalent and Thai's eat ~250 g per day, and found an intake of 80 μg/d inorganic arsenic (Ruangwises and Saipan 2010).

A study looking at major food groups contributing to arsenic intake in Korean population ignored rice because it ignored speciation (Lee et al. 2006), and while rice is low in total arsenic compared to seafoods, which are prevalent in the Korean diet, it has a high percentage of inorganic arsenic (Fig. 2.2). It was not that rice was not analysed in this study, rather that the dietary intakes were done for the ten foods highest in any particular element.

3.4 The Effects of Cooking Rice on Its Arsenic Content

Another variable that must be considered when determining rice exposure is cooking. Arsenic content of rice can either be increased, decreased or unchanged dependent on the method of rice cooking and the arsenic concentration of the cooking water. Of major concern is in regions of the world that suffer from elevated arsenic in household water supplies, such as Bangladesh and West Bengal, where there is already elevated arsenic in drinking water, rice cooking is a major contributory route to inorganic arsenic ingestion.

Bae et al. (2002) were the first to study the effects of cooking rice in arsenic contaminated water. Their study was based in Bangladesh using traditional cooking techniques. Market rice was boiled in an aluminium pot. The rice (500 g) had around 2.5 L of water added, with the rice absorbing around 1.2 L of this water. The water had 0.23–0.37 mg/L inorganic arsenic, causing the arsenic concentration of the cooked rice to rise from 0.17 to 0.21–0.31 mg/kg As after cooking. This elevation of inorganic arsenic in rice has also been demonstrated by others (Rahman et al. 2006; Sengupta et al. 2006; Smith et al. 2006). Ackerman et al. (2005) found that 89–105% of arsenic added in the cooking water was absorbed to the rice. A probabilistic model of an arsenic affected region of West Bengal found that rice cooking contributed 8% of dietary intake of inorganic arsenic (Mondal and Polya 2008).

The way in which rice is cooked affects its arsenic concentration. Modern cooking, either pressure cooking or using low volumes of water on open boiling so that cooking water is absorbed by the rice, retains arsenic content of the rice (Rahman et al. 2006; Raab et al. 2008). Raab et al. (2008) conducted systematic studies into arsenic speciation in rice following different forms of cooking and found arsenic speciation to be unchanged when rice was cooked in a volume of water that was absorbed by cooking. While disadvantageous with respect to arsenic content, this low volume method of cooking retains vitamins and nutrients, particularly soluble B vitamins, and thus there have been education campaigns to promote this form of cooking.

Traditionally in Asia, rice is thoroughly rinsed before cooking and then cooked with a large excess of water, with the water being drained of after cooking. Both rinsing and high volume of water cooking reduces arsenic concentration (Raab et al. 2008; Rahman et al. 2006; Sengupta et al. 2006), specifically the inorganic arsenic, by about 30%, but has no effect on DMA concentration (Raab et al. 2008). This specific removal of inorganic arsenic by high volume cooking is probably due to localization of arsenic species in rice grain. Inorganic species are elevated in the outer parts of the grain due to physiological reasons regarding grain unloading while DMA is more prevalent in the body of the endosperm (Carey et al. 2010; Meharg et al. 2008).

Sengupta et al. (2006) cooked a range of rice samples in water containing 0.003 mg/L arsenic using either traditional high water volume cooking technique or a low volume cooking technique. In the traditional method rice was first washed until the rinsing water was clear, which reduced the arsenic concentration of the rice by 28%. Rice was cooked in a 1:6 rice to water volume with excess water discarded. This cooking method removed 57% of the arsenic, including that lost in the rinse step. Using no rinse and a 1:1.5–2 L rice-water volume, the arsenic concentration of the rice stayed unchanged.

In summing up, where rice is cooked in a small volume of low arsenic water, the arsenic concentration and speciation of rice differs little between raw and cooked rice. Where large volumes of water are used in cooking, with discarding of the cooking water, rice arsenic levels decreases by up to 50%.

3.5 Human Bioavailability of Arsenic in Rice

The next stage in identifying the risk posed by inorganic arsenic in the diet is human gut bioavailability of that arsenic. Arsenic is relatively rapidly excreted from the body, primarily as urine (Casini et al. 2011). Therefore, urinary excretion of arsenicals is a good way to monitor bioavailability. There is only one, limited, rice eating trial where rice consumption was measured and urinary excretion monitored in humans. As the study included only two individuals, the results have to be interpreted with some caution. Survey of high and low rice eating communities, Bangladeshis and Caucasians, living in the UK found *circa* threefold elevated levels of arsenicals that derive from rice, including inorganic arsenic and its human metabolite, namely DMA (Casini et al. 2011). As Bangladeshi rice consumption is much greater than Caucasians (Table 3.2), and because elevated sources of inorganic arsenic and DMA into the human diet are minimal for most foods, besides rice, these findings can be interpreted as inorganic arsenic in rice being bioavailable.

There has only been limited study using animal models on the bioavailability of arsenic from rice. Juhasz et al. (2006) studied the bioavailability of arsenic species from rice in pigs, where blood samples were taken at different times following ingestion of rice or gavaged pure arsenic species. Orally gavaged pure arsenic salts differed considerably in their bioavailability, with arsenite having 104%, arsenate 94%, DMA 33% and MMA 17% bioavailability. When rice high in DMA (86%) was fed to pigs only 33% was bioavailable, whereas inorganic arsenic in rice was 89% bioavailable. They concluded that inorganic arsenic from rice was highly gut available, while DMA was poorly available.

In vitro bioavailability studies are more common in the literature, where rice is subjected to simulated gut enzymatic and chemical conditions. Laparra et al. (2005) lyophilised rice cooked in arsenic contaminated water and subjected it to pepsin and then to pancreatin and bile extract. The soluble extract was then exposed to a monolayer of immobilized Caco-2 cells, cultured human colon cells, seeded onto a polycarbonate surface. The bioaccessible fraction varied from 83% to 120%, showing that both the bioincorporated and the additional dosed arsenic were highly bioaccessible. However, only 4–18% of this bioaccessible arsenic, which was primarily inorganic due to the manner in which it was dosed, was assimilated by the Caco-2 cells. This contrasts strongly with the *in vivo* bioavailability of arsenic to swine (Juhasz et al. 2006), where inorganic arsenic dosed into rice via cooking was 89% mobilised into the bloodstream. This suggests that the Caco-2 cell model is a poor predictor of bioavailability. A very similar simulated enzymatic approach to Laparra et al. was used by Ackerman et al. (2005) to look at the bioaccessible arsenic from rice cooked in arsenic free water. They found that the enzymatic extraction liberated, on average, 94% arsenic. *In vitro* studies have major interpretational limitations (Ackerman et al. 2005; Laparra et al. 2005), but the potential for arsenic in rice grain to be mobilized from the gut is apparent.

Given our current understanding, when conducting risk assessments, given that such assessments should ere on the side of caution in the absence of concrete parameterization, it can be assumed that the availability of inorganic arsenic from rice in the gut is 100%. When accurate data are available, any risk assessment models can subsequently incorporate more refined parameterization.

3.6 Risk Assessment from Arsenic in Rice

To date, no epidemiological study published has attempted to define the risks from ingesting inorganic arsenic in rice. As it is only a decade or so since inorganic arsenic was identified to be specifically elevated in rice, and even shorter time since it was realised, and then widely accepted, that rice constitutes a major pathway of inorganic arsenic into the human diet (EFSA 2009), this lack of epidemiological knowledge is hardly surprising. Furthermore, even if epidemiological studies were to be initiated, it would take decades to understand how elevated arsenic in rice affects lifetime health outcomes. Most of the diets rich in arsenic are in SE Asia which has historically had low average life expectancies, low rates of medical intervention, and poor health record keeping. Cancers that effect a populace in later life have not been well characterised in this region. However, with life expectancy constantly increasing, along with wealth, late lifetime cancers will come to the fore. It is only recently, after five decades of research, that there is an understanding of low level risks posed by arsenic in drinking water starting to be reached, that is below 50 μl/L inorganic arsenic (NRC 2001; WHO 2004), and much of that research focuses on molecular markers rather than direct illness, given the inherent problems of following through epidemiological studies for decades for a chronic carcinogen, such as inorganic arsenic. Eating 0.5 kg of rice with 0.1 mg/kg arsenic equates to drinking 1 L of water containing 50 μl/L arsenic. In the absence of epidemiological data collected with respect to exposure to inorganic arsenic from rice, or indeed equivalent exposures from drinking water, risks posed from arsenic in rice have to be extrapolated from high arsenic in drinking water scenarios, as has traditionally been the case for low arsenic drinking water scenarios, and indeed US, EU and WHO drinking water standards were set using this extrapolation (Meharg and Raab 2010). As the human body is unable to distinguish if inorganic arsenic in the bloodstream is from water or rice, the risks posed by inorganic arsenic in rice can be considered the same as that from waters, with the qualification that water will have higher, but only slightly (see Sect. 3.5), bioavailability than inorganic arsenic in rice.

The US National Research Council (NRC) reviewed risks posed by arsenic in drinking water and established that the inorganic arsenic dose response curves for lung and bladder cancers from drinking water were linear (NRC 2001). Theoretical maximum-likelihood estimates of excess lifetime risk for inorganic arsenic, expressed as an incidence per 10,000 people were calculated based on consumption of 1 L of water per day, based on Taiwanese epidemiological studies. This dose response modelling was used as the scientific rational of reducing arsenic in US drinking water from the old standard of 0.05–0.01 mg/L. Cancer slopes for arsenic are under constant review. The current US EPA skin cancer slope is 1.5 per mg As/kg body mass/d excess lifetime risk for inorganic arsenic, but a slope of 3.67 per mg/kg/d has been used in recent US EPA assessments (Tsuji et al. 2007). It is important to note that these risks are of the same order of magnitude, and given the variance associated with lifetime cancer risks, and indeed that inorganic arsenic gives rise to a range of cancers (balder, lung and skin to the foremost), as well as a range of other potentially confounding health effects (NRC 2001), such generic risk slopes can never be anything else but ballpark figures. Again, until detailed epidemiological study, taking into account a wide range of factors that may interact with lifetime cancer risks, such as smoking, nutrition, occupation, genetic predisposition etc., determining the variance on cancer rates, are conducted, such generic slope figures are the only pragmatic way to proceed in estimating risks.

The EPA has an upper limit of acceptable risk for cancer from any given source of 1 in 10,000 (Tsuji et al. 2007). This is a useful figure with which to consider cancer risks from inorganic arsenic from rice. For the US populace, Tsuji et al. (2007) calculated that at the 95th percentile a 6.1 μg/d inorganic arsenic ingestion rate for rice, at a slope of 1.5 per mg/kg/d, for a 65 kg person, equates to an excess skin cancer risk of 1.4 in 10,000. Using the slope of 3.67 per mg/kg/d, this equates to an excess cancer rate of 3.4 in 10,000. The mean arsenic ingestion rate from rice from Tsuji et al. (2007) is 1.67 μg/d, i.e. 27% of the 95th percentile. Using a slope of 3.67 per mg/kg/d the excess skin cancer rate from mean levels of inorganic arsenic ingestion from rice is 0.9 in 10,000. The situation is more problematic for children (1–6 y) with a mean and 95th percentile inorganic arsenic intake from rice of 1 and 3.1 μg/d, respectively, assuming an average body mass of 15 kg (Tao and Bolger 1999).

If UK Bangladeshi adults, the highest UK rice consuming group, are considered, for a 65 kg person consuming 0.25 kg of rice per day (Table 3.2), assuming an inorganic arsenic concentration in rice of 0.1 mg/kg (Sect. 2.3), the excess cancer risk would be 5.8 and 14.1 per 10,000 based on a slope of 1.5 and 3.67 per mg/kg/d, respectively. For other UK consumers, assuming consumption of rice with an inorganic arsenic concentration of 0.1 mg/kg, based on a 65 kg body mass, the relative cancer rate can be calculated. For example, for white ethnic origin, based on ingestion of 8 g of rice per day, an excess cancer risk of 0.2 and 0.5 in 10,000 can be calculated for a 1.5 and 3.67 per mg/kg/d slope, respectively.

Mondal and Polya (2008) probabilistically modelled arsenic cancer risks for an arsenic affected region of West Bengal and found that a median increased lifetime cancer risk of 7.2 per 10,000 was attributable to eating rice. This compares to 22 per 10,000 calculated through empirical modelling by Meharg et al. (2009) for neighbouring Bangladesh (Table 3.4).

3.7 Food Arsenic Standards

Arsenic is a chronic carcinogen, as well as an acute toxin at higher concentrations, and decades of exposure to elevated levels lead to a host of illnesses including bronchitis, hypertension, miscarriage, skin hypo and hyper pigmentation, skin, bladder and lung cancers (Smith et al. 1992, 2006; WHO 2004). It is the inorganic forms of arsenic that are thought to be of concern with respect to these illnesses, but evidence suggests that organic forms, including DMA, should receive more attention (Borak and Hosgood 2007; Smith et al. 2006; Yakanaka et al. 2004). This is because it is thought potentially that in the human body that DMA(V) can be reduced to DMA(III) by compounds such as glutathione or enzymatically via arsenate reductases. Like inorganic arsenic, it is the reduced species, or at least the redox cycling between the reduced and oxidised state, that is thought to give rise to carcinogenicity (WHO 2004). However, as is standard in the literature, and because no chronic dose response curves have been established for DMA, only exposure to inorganic arsenic will be considered.

Food arsenic standards are notable by their absence with only some South East Asian countries, China to the fore (Table 3.5) having set standards in modern times. The Chinese set inorganic arsenic rice standard 0.15 mg/kg (USDA 2006). From the US rice survey of Williams et al. (2007) 20% of US rice would fail this standard. From the rice arsenic speciation data in the basket survey presented in Fig. 2.2, it is

Table 3.4 Excess lifetime cancer risks for rice consumption by country

	Polished rice consumption (g/d)	Median inorganic As content of rice (mg/kg)	Country specific rice inorganic As intake (µg/d)	100 g per day rice inorganic As intake (µg/d)	Country specific rice excess cancer rate per 10,000	100 g per day rice excess cancer rate per 10,000
Bangladesh	445	0.081	36.2	8.1	22.1	5.0
China	218	0.109	23.7	10.9	15.2	7.0
India	192	0.059	11.3	5.9	6.9	3.6
Italy	17	0.071	1.2	7.1	0.7	4.3
USA	24	0.088	2.1	8.8	1.3	5.4

Data are from Meharg et al. (2009)

Table 3.5 Standards for the maximum levels of arsenic in foods, Peoples Republic of China (USDA 2006)

Item	Total arsenic (mg/kg)	Inorganic arsenic (mg/kg)
Rice		0.15
Flour		0.1
Other grains		0.2
Vegetables		0.05
Fruits		0.05
Poultry and meat		0.05
Eggs		0.05
Milk powder		0.25
Fresh milk		0.05
Beans		0.1
Wines		0.05
Fish		0.1
Algae (d. wt.)		1.5
Shellfish and crustaceans (f. wt.)		0.5
Shellfish and crustaceans (d. wt.)		1.0
Other aquatic products		0.5
Edible oil and fat	0.1	
Fruit juice and syrup	0.2	
Cacao fat and chocolate	0.5	
Other cacao products	1.0	
Sugar	1.0	

evident that a substantial proportion of Chinese, Italian, Spanish and US rice would fail this standard. It should be noted, however, that the rational for setting this standard was not outlined, so the utility of this threshold is uncertain.

In the absence of modern food standards, old regulations are evoked. For instance, in the absence of EU standards, which regulate food safety for member states, the UK has only a 1959 standard of 1 mg/kg total arsenic to fall back on (HMSO 1959). Based on ~50% of arsenic in rice being inorganic (Fig. 2.2), this would set the "maximum" level of inorganic arsenic permissible in rice to be ~0.5 mg/kg. This is clearly highly inappropriate. Furthermore, if the logic of this standard setting was taken to its conclusion then most seafood would be banned for consumption as they have high total arsenic, but mostly in the form of relatively non-toxic species such as arsenobetaine and arsenosugars (Francesconi 2007).

The wider fallback position for considering inorganic arsenic from foodstuffs is the World Health Organization's Permissible Maximum Tolerable Daily Intake (WHO PMTDI) of 2.1 μg/kg/d. This PMTDI allows approximately tenfold higher, dependent on body mass of the consumer, intake of inorganic arsenic than assumed under EU and US regulations for water set at 10 μl/L, assuming ingestion of 1 L of water per day for a 50 kg person. This juxtaposition between water standards and PMTDI needs to be resolved (Meharg and Raab 2010) and PMTDI has come into criticism for not being set with a firm scientific basis (EFSA 2009). This EFSA criticism was closely followed by the WHO revoking the PMTDI, but at the time of writing no alternatives are in place, effectively meaning that there are even fewer standards to judge thresholds for arsenic in rice. Both the EU and WHO are actively pursuing standard settings for inorganic arsenic in foodstuffs so hopefully the situation will soon be rectified and meaningful standards based on adequate risk assessments put into place.

3.8 Limiting Arsenic Exposure to Rice

Much of the focus of this book is on understanding the dynamics of arsenic exposure in rice agroecosystems, with the ultimate aim of reducing inorganic arsenic concentration of rice, through paddy management and rice genotype selection. For cultures that produce and consume large quantities of rice, reducing arsenic concentration of locally or regionally produced rice is the only viable option in terms of economics and logistics (Meharg and Raab 2010). For those countries whose sole or major sources of rice are through imports, there are further options, namely by altering import patterns or switching to other carbohydrate sources. This option of sourcing rice to reduce dietary arsenic is relevant to the theme of this chapter. The subject of rice sourcing to reduce arsenic in the diet does not need much expansion as the logic is simple (Meharg and Raab 2010), if various choices as to where to source rice from are available, then that at the lower range of inorganic arsenic can be selected.

A range of factors may complicate such selection such as trade agreements, export embargoes and price amongst them. It is primarily non-traditional rice eating cultures of developed countries not capable, because of climatic constraints, of producing their own rice that have the most choices with respect to where rice is imported from. Such countries tend to be the lowest rice consumers (Fig. 3.1) and at average rates of rice consumption for these countries, even if the rice is at the upper end of the inorganic arsenic concentration of market available rice, there is little excess lifetime cancer risk so there may be little need in altering current rice import patterns for the bulk populace of such countries. More pertinent is to consider high rice consumers discussed elsewhere in the chapter, such as ethnic sub-populations, those on specialised health or life diets, and young children, who either consume high amounts of rice and/or have large food intakes with respect to their body mass. High arsenic in baby rice has been raised as a particular issue (Meharg et al. 2008), and subsequently identified as a cause of concern by EFSA (EFSA 2009). It would be simple to reduce the total arsenic concentration of baby rice by ~4-fold by sourcing rice from low arsenic regions of the world (Fig. 2.2). Figure 3.3 shows the balance of production, import and export for each country on the FAO rice database and it can be seen that it is the non- or low production rice growing nations that have the largest flexibility in controlling the quality of rice they import.

Another strategy is to use alternative carbohydrate sources but, again, this is really only viable in countries that do not have a reliance on rice. Those countries

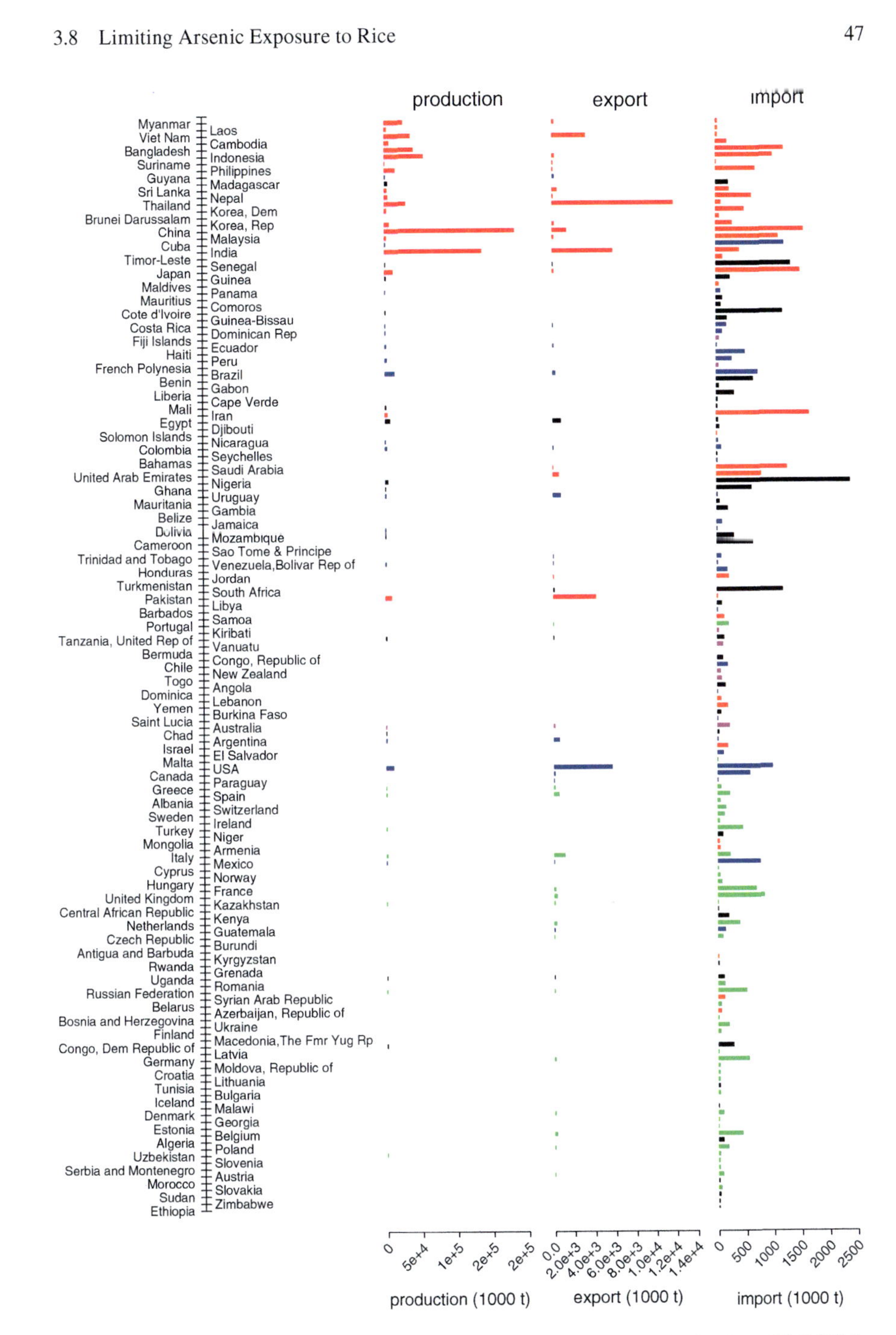

Fig. 3.3 Rice production, import and export by country compiled from FAO data base (FAO 2004)

with a reliance on rice, therefore, must consider agronomic or crop breeding strategies for reducing grain arsenic, discussed in Chap. 7.

References

Ackerman AH, Creed PA, Parks AN, Fricke MW, Schwegel CA, Creed JT, Heitkemper DT, Vela NP (2005) Comparison of a chemical and enzymatic extraction of arsenic from rice and an assessment of the arsenic absorption from contaminated water by cooked rice. Environ Sci Technol 39:5241–5246

Agusa T, Kunito T, Minh TB, Trang PTK, Iwata H, Viet PH, Tanabe S (2009) Relationship of urinary metabolites to intake in residents of the Red River Delta, Vietnam. Environ Pollut 157:396–403

Argos M, Karla T, Rathouz PJ, Chen Y, Pierce B, Parvez F, Islam T, Ahmed A, Rakibuz-Zaman M, Hasan R, Sawar G, Slavkovich V, van Geen A, Graziano J, Ahsan H (2010) Arsenic exposure from drinking water, and all-cause and chronic-disease mortalities in Bangladesh (HEALS): a prospective cohort study. Lancet 376:252–258

Bae M, Watanabe C, Inaoka T, Sekiyama M, Sudo N, Bokul MH, Ohtsuka R (2002) Arsenic in cooked rice in Bangladesh. Lancet 360:1839–1840

Batres-Marquez SP, Jensen HH (2005) Rice consumption in the United States: new evidence from food consumption surveys. Food and Nutrition Policy Division, Centre for Agricultural and Rural Development, Iowa State University. Staff Report 05-SR 100

Borak J, Hosgood HD (2007) Seafood arsenic: implications for human risk assessment. Regul Toxicol Pharmacol 47:204–212

Carey AM, Scheckel KG, Lombi E, Newville M, Choi Y, Norton GJ, Charnock JM, Feldmann J, Price AH, Meharg AA (2010) Grain unloading of arsenic species in rice (*Oryza sativa* L.). Plant Physiol 150:309–319

Casini C, Raab A, Jenkins RO, Feldmann F, Meharg AA, Haris P (2011) The impact of a rice based diet on urinary arsenic. J Environ Monit 13:257–265

EFSA (2009) Panel on Contaminants in the Food Chain (CONTAM)/Scientific opinion on arsenic in food. EFSA J 7:1351

FAO (Food and Agriculture Organization of the United Nations) (2004) http://faostat.fao.org/site/336/DesktopDefault.aspx?PageID=336

Francesconi KA (2007) Toxic metal species and food-regulations – making a healthy choice. Analyst 132:17–20

HMSO (1959) The arsenic in food regulations. London

Juhasz AL, Smith E, Weber J, Rees M, Rofe A, Kuchel T, Sansom L, Naidu R (2006) In vivo assessment of arsenic bioavailability in rice and its significance for human health risk assessment. Environ Health Perspect 114:1826–1831

Kile ML, Houseman EA, Breton CV, Smith T, Quamruzzaman O et al (2007) Dietary arsenic exposure in Bangladesh. Environ Health Persp 115:889–893

Lamont WH (2003) Concentration of inorganic arsenic in samples of white rice from the United States. J Food Compost Anal 16:689–695

Laparra JM, Velez D, Barbera R, Farre R, Montoro R (2005) Bioavailability of inorganic arsenic in cooked rice: practical aspects for human health risk assessments. J Agric Food Chem 53:8829–8833

Lee HS, Cho YH, Park SO, Kye SH, Kim BH, Hahm TS, Kim M, Lee JO, Kim C (2006) Dietary exposure of the Korean population to arsenic, cadmium, lead and mercury. J Food Compost Anal 19:31–37

Lin HT, Wong SS, Li GC (2004) Heavy metal content of rice and shellfish in Taiwan. J Food Drug Anal 12:167–174

Liu H, Probst A, Liao B (2005) Metal contamination of soils and crops affected by the Chenshou lead/zinc mine spill (Hunan, China). Sci Total Environ 339:153–166

Meacher DM, Menzel DB, Dillencourt MD, Bic LF, Schoof RA, Yost LJ, Eickhoff JC, Farr CH (2002) Estimation of multimedia inorganic arsenic intake in the US population. Hum Ecol Risk Assess 8:1697–1721

Meharg AA (2007) Arsenic in rice – a literature review. Food Standards Agency, contract C101045

Meharg AA, Raab A (2010) Getting to the bottom of arsenic standards and guidelines. Environ Sci Technol 44:4395–4399

Meharg AA, Williams PN, Schekel K, Lombi E, Feldmann J, Raab A, Zhu YG, Gault A, Islam R (2008) Speciation of arsenic differs between white and brown rice grain. Environ Sci Technol 42:1051–1057

Meharg AA, Williams PN, Adamako E, Lawgali YY, Deacon C, Villada A, Cambell RCJ, Sun GX, Zhu YG, Feldmann J, Raab A, Zhao FJ, Islam R, Hossain S, Yanai J (2009) Geographical variation in total and inorganic arsenic content of polished (white) rice. Environ Sci Technol 43:1612–1617

Meliker JR, Franzblau A, Slotnick MJ, Nriagu O (2006) Major contributors to inorganic arsenic intake in southeastern Michigan. Int J Hyg Environ Health 209:399–411

Mohri T, Hisanaga A, Ishinishi N (1990) Arsenic intake and excretion by Japanese adults: a 7-day duplicate diet study. Food Chem Toxicol 28:521–529

Mondal D, Polya DA (2008) Rice is a major exposure route for arsenic in Chakdaha block, Nadia district, West Bengal, India: a probabilistic risk assessment. Appl Geochem 23:2987–2998

National Diet and Nutrition Survey (1992) Toddlers aged 1.5 to 4.5 years. http://www.food.gov.uk/multimedia/pdfs/publication/ndnsreport0809.pdf, accessed 02/01/2012

National Diet and Nutrition Survey (1997) Young people aged 4 to 18 years. http://www.food.gov.uk/multimedia/pdfs/publication/ndnsreport0809.pdf, accessed 02/01/2012

National Diet and Nutrition Survey (2000) Adults aged 19 to 64 years. http://www.food.gov.uk/multimedia/pdfs/publication/ndnsreport0809.pdf, accessed 02/01/2012

National Research Council (2001) Arsenic in drinking water – 2001 update. National Academy Press, Washington, D.C

Navas-Acien A, Francesconi KA, Silbergeld EK, Guallar E (2011) Seafood intake and urine concentrations of total arsenic, dimethlarinate and arsenobetaine in the US population. Environ Res 111:110–118

Ohno K, Yanase T, Matsuo Y, Kimura T, Rahman MH et al (2007) Arsenic intake via water and food by a population living in an arsenic-affected area of Bangladesh. Sci Total Environ 381:68–76

Raab A, Baskaran C, Feldmann J, Meharg AA (2008) Cooking rice in a high water to rice ratio reduces inorganic arsenic content. J Environ Monit 11:41–44

Rahman MA, Hasegawa H, Rahman MA, Rahman MM, Miah MAM (2006) Influence of cooking method on arsenic retention in cooked rice related to dietary exposure. Sci Total Environ 370:51–60

Rahman AM, Hasegawa H, Rahman MM, Miah MMA, Tasmin A (2008) Arsenic accumulation in rice (*Oryza sativa* L.): human exposure through food chain. Ecotoxicol Environ Saf 69:317–324

Ruangwises S, Saipan P (2010) Dietary intake of total and inorganic arsenic by adults in an arsenic contaminated area of Ron Phibun District, Thailand. Bull Environ Contam Toxicol 84:274–277

Schoof RA, Yost LJ, Eickhoff J, Crecelius EA, Cragin DW et al (1999) A market basket survey of inorganic arsenic in food. Food Chem Toxicol 37:839–846

Sengupta MK, Hossain MA, Mukherjee A, Ahamed S, Das B, Nayak B, Pal A, Chakraborti D (2006) Arsenic burden of cooked rice: traditional and modern methods. Food Chem Toxicol 44:1823–1829

Signes-Pastor AJ, Mitra K, Sarkhel S, Hobbes M, Burlo F, De Groot WT, Carbonell-Barrachina AA (2008) Arsenic speciationin food and estimation of dietary intake of arsenic in a rural village of West Bengal, India. J Agric Food Chem 56:9469–9474

Smedley PL, Kinniburgh DG (2002) A review of the source, behaviour and distribution of arsenic in natural waters. Appl Geochem 17:517–568

Smith AH, Hopenhaynrich C, Bates MN, Goeden HM, Hertzpiccioto I, Duggan HM, Wood R, Kosnett MJ, Smith MT (1992) Cancer risks from arsenic in drinking water. Environ Health Pers 97"259:267

Smith NM, Lee R, Heitkemper DT, Cafferty KD, Haque A, Henderson AK (2006) Inorganic arsenic in cooked rice and vegetables from Bangladeshi households. Sci Total Environ 370:294–301

Tao SSH, Bolger PM (1999) Dietary arsenic intakes in the United States: FDA total diet study, September 1991-December 1996. Food Addit Contam 16:465

Thompson T (2001) Wheat starch, gliadin, and the gluten-free diet. J Am Diet Assoc 101:1456–1459

Tsuji JS, Yost LJ, Barraj LM, Scrafford CG, Mink PJ (2007) Use of background inorganic arsenic exposures to provide perspective on risk assessment results. Regul Toxicol Pharmacol 48:59–68

Uchino T, Roychowdhury T, Ando M, Tokunaga H (2006) Intake of arsenic from water, food composites and extretion through urine, hair from a studied population in West Bengal, India. Food Chem Toxicol 44:455–461

USDA (2006) Foreign Agricultural Service Global Agriculture Information Network Report CH6064. Peoples Republic of China, FAIRS Product Specific Maximum Levels of Contaminants in Foods

Wharton PA, Eaton PM, Wharton BA (1984) Subvariation in the diets of Moslem, Sikh and Hindu pregnant women at Sorrento Maternity Hospital, Birmingham. Br J Nutr 52:469–476

Williams PN, Price AH, Raab A, Hossain SA, Feldmann J, Meharg AA (2005) Variation in arsenic speciation and concentration in paddy rice related to dietary exposure. Environ Sci Technol 39:5531–5540

Williams PN, Raab A, Feldmann J, Meharg AA (2007) Market basket survey shows elevated levels of as in South Central US processed rice compared to California: consequences for human dietary exposure. Environ Sci Technol 41:2178–2183

World Health Organization (2004) IARC, Working Group on some drinking water disinfectants and contaminants, including arsenic, vol 84. Lyon. Monograph 1

Yakanaka K, Kato K, Mizoi M, An Y, Takabayashi F, Nakano M, Hoshino M, Okada S (2004) The role of arsenic species produced by metabolic reduction of dimethylarsenic acid in genotoxicity and tumorigenesis. Toxicol Appl Pharm 198:385–393

Yost LJ, Tao SH, Egan SH, Barraj LM, Smith KM, Tsuji JS, Lowney YW, Schoof RA, Rachman NJ (2004) Estimation of dietary intake of inorganic arsenic in US children. Human Ecolog Risk Asses 10:473–483

Zavala YJ, Gerads R, Gürleyük H, Duxbury JM (2008) Arsenic in rice: II. Arsenic speciation in USA grain and implications for human health. Environ Sci Technol 42:3861–3866

Chapter 4
Sources and Losses of Arsenic to Paddy Fields

4.1 Introduction

Arsenic in rice is derived from the soil matrix that sustains it. Paddy field biogeochemistry leads to excessive arsenic mobilization and subsequent assimilation by rice (see Chap. 5), though arsenic has to be present in the first place to be mobilized, which will be the subject of the present chapter. While a consideration of the inputs and sources of arsenic into paddies is important, so are potential losses from the soils, and it is becoming apparent that a mass balance approach is required (Dittmar et al. 2010; Neumann et al. 2011; Roberts et al. 2010). Furthermore, it is starting to be realised that in some agro-ecosystems, particularly the groundwater irrigated paddies of SE Asia, that rice cultivation may greatly alter wider arsenic biogeochemical cycles through the surface to groundwater continuum, potentially affecting groundwater arsenic (Burgess et al. 2010; Neumann et al. 2010). This topic will also be considered here.

4.2 Sources of Arsenic to Paddy Ecosystems

Arsenic sources to paddies can be divided into natural and anthropogenic. Natural inputs are sub-divided into inherent soil arsenic, and any arsenic carried in additionally through flooding (aqueous and sediment), and wet and dry atmospheric deposition.

Anthropogenic sources are multitudinous and highly variable and can be considered in the following classes:

(a) non-point source industrial/urban pollution for paddies downstream of large population centres;
(b) use of fertilizers and organic manures contaminated with arsenic;
(c) point-source industrial pollution;
(d) use of arsenical pesticides;
(e) contamination of irrigation water.

A.A. Meharg and F.J. Zhao, *Arsenic & Rice*, DOI 10.1007/978-94-007-2947-6_4,

Table 4.1 Arsenic concentrations in geological materials

Medium	Concentration (mg kg^{-1})	Source
Universe	0.008	Matschullat (2000)
Sun	0.004	Matschullat (2000)
Stony meteorites	1.8	Matschullat (2000)
Iron meteorites	11	Matschullat (2000)
Earth crust total	1–1.8	Matschullat (2000)
Upper crust	1.5–2	Matschullat (2000)
Utramafic rocks	0.7	Matschullat (2000)
Mafic igneous (mainly dolerite and basalt)	0.2–113	Hale (1981)
Oceanic ridge basalts	1	Matschullat (2000)
Gabbros, basalts	0.7	Matschullat (2000)
Granites, granodiorites	3	Matschullat (2000)
Acid igneous (mainly granite)	0.2–15	Hale (1981)
Sandstones	0.5–1	Matschullat (2000)
Shales, schists	13	Matschullat (2000)
Shale and argillite	0.3–500	Hale (1981)
Schist and phyllite	0.5–143	Hale (1981)
Carbonates	1–1.5	Matschullat (2000)
Metamorphites	0.1–11	Matschullat (2000)
Minerals		
Pyrite	5–5,600	Hale (1981)
Chalcopyrite	10–1,000	Hale (1981)
Galena	5–60	Hale (1981)
Sphalerite	5–400	Hale (1981)

The last 3 (c-e) are normally considered to be of most concern. The following sections will systematically consider natural and anthropogenic sources by the generic classes identified.

4.2.1 Natural Sources

Arsenic levels in the earth's crust are below 2 mg kg^{-1}, but can be much higher in igneous and metamorphic rocks, and lower in sandstones and carbonates (Table 4.1). Arsenic levels can be naturally highly elevated in zones of active or extinct volcanic activity, particularly in hot-spring environments (Smedley and Kinniburgh 2002).

A wide range of minerals (>200) contain arsenic (Smedley and Kinniburgh 2002), and where these minerals outcrop, their weathering can lead to naturally high levels in groundwaters, surface waters and soils. Examples of arsenic levels in some minerals are given in Table 4.1. Arsenic is often found to be concentrated in sulphur bearing minerals, and has a strong affinity for pyrite, a frequently occurring mineral (Nordstrom 2002). Arsenic can substitute for sulphur in the mineral lattice. Arsenic is also concentrated on iron hydroxides (Smedley and Kinniburgh 2002). Arsenic can be released from both pyrites and iron hydroxides under certain redox conditions.

Koljonen (1992) estimates a global average of arsenic in soils of 5 mg kg^{-1}. Uncontaminated (i.e. no additional anthropogenic inputs) soils of the US range from 4.5 to 13 mg kg^{-1}, with an average of 7.5 mg kg^{-1} (Goldschmidt 1958). Allard (1995) gives this range for the US as 0.1–55 mg kg^{-1}, with an average value of 7.2 mg kg^{-1}. A limited range of geological environments can result in significant natural elevation of arsenic in environmental materials (Nordstrom 2002). These include: organic rich (black) shales, Holocene alluvial sediments, mineralized zones, volcanogenic sources, thermal springs, closed basins in arid-to-semiarid climates, particularly in volcanic regions, strongly reducing aquifers with low sulphate concentrations (WHO 2004).

Soils formed from arsenic enriched geological substrates can have naturally much higher arsenic levels than the ranges quoted above. These "normal" ranges must therefore, be considered as typical background levels, rather than absolute ranges for natural soil arsenic as soils formed on top of arsenic rich bedrocks will be elevated in this element. Colbourn et al. (1975) report mean arsenic levels of 88 mg kg^{-1}, with a range of 24–250 mg kg^{-1} (n = 18), in soils formed naturally from parent material consisting of metamorphic aureole around a granitic intrusion. Soils formed, for example on arsenopyrite (FeAsS) mineralization, with the ore body running close to the surface have arsenic levels ranging from 700 to 4,000 mg kg^{-1} (Geiszinger et al. 2002). Soils formed in and around ancient and modern hot-springs with elevated arsenic in geothermal fluids will have naturally elevated levels of arsenic due to enrichment of soil parent material (Ballantyne and Moore 1988).

Depending on prevailing climatic and hydrological conditions, soils and sediments, surface waters, groundwaters and air can become enriched in arsenic where these geological conditions prevail. Some of the major rice growing regions of the world suffer such natural elevation. The best example of natural elevation is the Holocene tracts of Bengal Basin which are higher in arsenic than those of the Pleistocene terraces of this region. Lu et al. (2009) studied ten Holocene and ten Pleistocene paddies dispersed over Bangladesh and found that the Holocene soils had approximately five-fold more total arsenic than the Pleistocene soils. Williams et al. (2011) surveyed a wide selection of Holocene and Pleistocene and found the same trend. The explanation for the higher Holocene levels is thought to be simply weathering, with the Pleistocene soils being more weathered, and thus having a lower readily weatherable mineral content in general (Williams et al. 2011). A similar situation is found for groundwaters from Holocene verses Pleistocene sediment (Smedley and Kinniburgh 2002). This difference has major consequences for arsenic levels in Bangladeshi rice (Lu et al. 2009; Williams et al. 2011). The widespread presence of rice cultivation on Holocene sediments in Bangladesh, in general, explains why Bangladesh has such a high baseline level of arsenic in rice grain (Lu et al. 2009; Zavala and Duxbury 2008). Problems of natural elevation of arsenic in Holocene sediments may also occur in other parts of the world. Winkel et al. (2008) used GIS approaches to map potential arsenic contamination in groundwater in SE Asia by mapping the distribution of Holocene sediments. The same approach may be appropriate to rice.

Organic rich sediments in SE Asia (Fig. 4.1) may also lead to arsenic mobilization due to reducing conditions as observed for the groundwaters of Taiwan (Schoof et al. 1998).

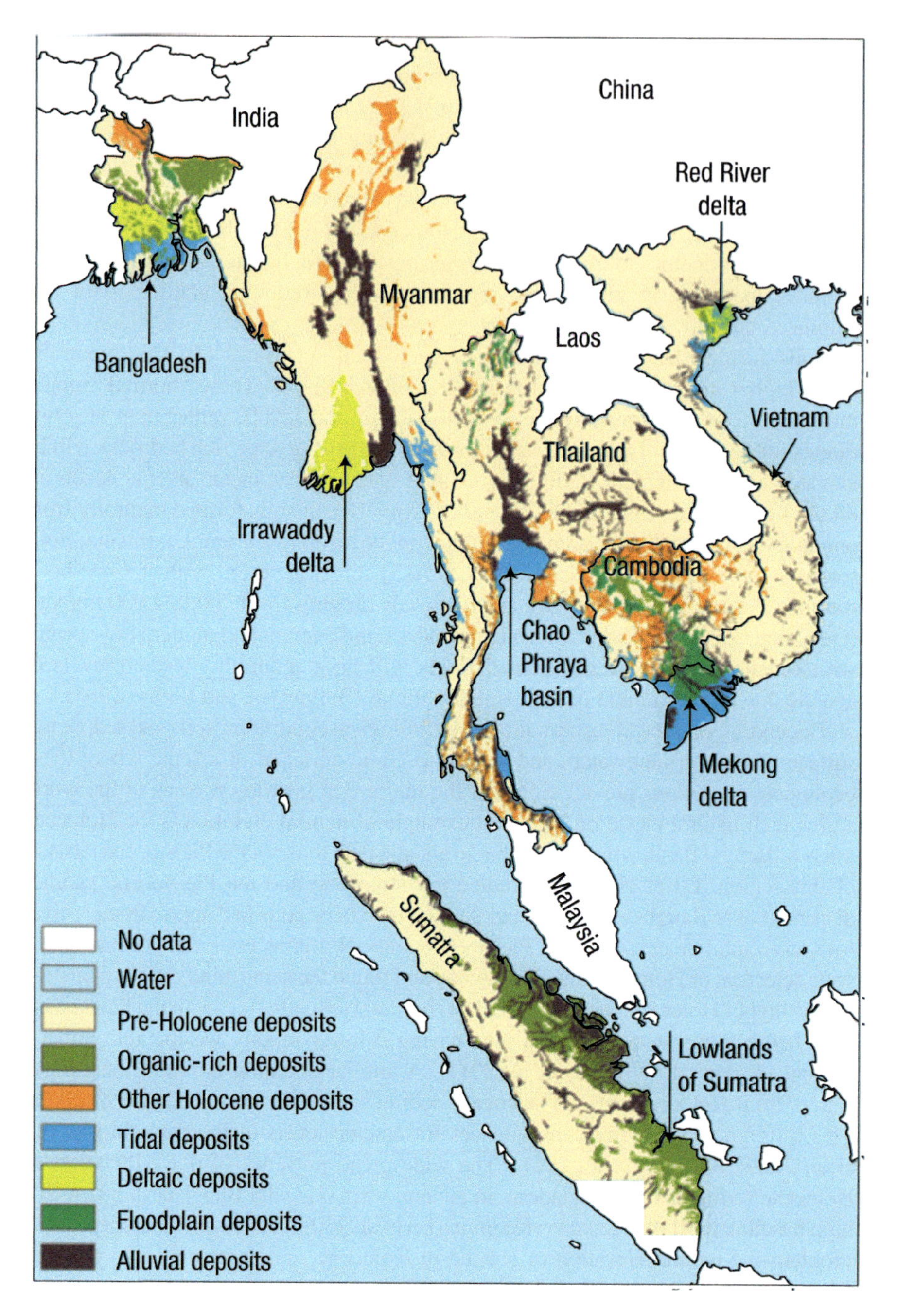

Fig. 4.1 Sediment distribution in SE Asia (Reprint with permission from Winkel et al. (2008))

Schoof et al. (1998) looked at potential arsenic contamination of rice in the affected regions of Taiwan, but the sampling size and structure was too limited to draw any conclusions. Japanese soils are elevated in arsenic along a spine associated with geothermal activity (Arao et al. 2009). Williams et al. (2007b) postulated that soils of the rice growing regions of California were naturally more elevated in arsenic than those of the South Central US. Whether such natural elevation in organic carbon in soils leads to elevated rice grain arsenic has yet to be determined.

While geothermal activity often leads to natural soil arsenic contamination, much of the arsenic will probably be locked away in non-bioaccessible forms. This is what has been observed for mining impacted paddy soils where there has been a much greater research focus with respect to rice contamination with arsenic (Norton et al. 2010; Williams et al. 2009; Zhu et al. 2008).

Ultimately, the bulk of all paddy soil arsenic will be of geogenic weathering origin, either due to soil formation from local bedrock, or from sediment carried in upstream. Atmospheric deposition is unlikely to be a major source of arsenic (Andreae 1980; Shotyk et al. 1996) except in industrialized zones (Clayton et al. 1999; Hubert et al. 1981).

4.2.2 *Non-point Source Industrial/Urban Pollution*

Agronomic areas are usually, though not exclusively, distant to industrial/urban areas where large-scale pollution can arise. This does not mean that industries and cities cannot provide direct routes of contamination to zones where rural agricultural is conducted, such as drainage channels and rivers that receive industrial/urban effluent, or where former industrial areas are latterly used for agriculture, or where contaminated soil/sediment is intentionally or accidently introduced into agro-ecosystems, or where diffuse atmospheric pollution leads to elevated soil pollution. Point source contamination is more readily understood, with numerous examples of arsenic impact on paddy fields (see Sect. 4.2.4). While diffuse sources of arsenic have not been studied, their role cannot be dismissed. Some major rice growing regions are downstream of industrial/urban centres such as the Camargue growing region of France, which has the average highest arsenic levels of any commercially purchased rice measured to date (Meharg et al. 2009). A similar scenario of diffuse industrial inputs has been discussed in the context of the rice growing region of the Ebro Valley, Spain (Ferre-Huguet et al. 2008).

4.2.3 *Fertilizers and Manures Contaminated with Arsenic*

Arsenic may be elevated in chemical fertilizers (Charter et al. 1995; Macedo et al. 2009), and fertilizer use in long term can lead to arsenic enrichment of croplands (Chen et al. 2008). Levels of arsenic found in globally sourced phosphate based fertilizers typically

Table 4.2 Arsenic concentrations in phosphate fertilizers sourced globally (From Charter et al. 1995)

Fertilizer type	n	Median (mg kg^{-1})	Minimum (mg kg^{-1})	Maximum (mg kg^{-1})
Triple super phosphate	24	10.1	2.4	18.5
Mononammonium phosphate	23	12.4	8.1	17.8
Diammonium phosphate	25	12.4	6.8	15.6
Rock phosphate	12	9.6	3.2	32.1

ranged from 3 to 30 mg kg^{-1} (Table 4.2). At a field application rate, typically 50 kg P_2O_5 per ha, such additions can only add a very small quantity of arsenic to soil. Assuming a soil bulk density of 1 and a fertilization depth of 10 cm, this equates to adding about 20 kg P to 1,000 tonnes of soil, or 20 mg P/kilogram soil. Assuming 10 mg As kg^{-1} in that fertilizer containing 20% P (typical of triple superphosphate), then this rate of fertilization would only add 1 μg As kg^{-1} soil per year. This yearly increment in arsenic addition from P fertilizers is small compared with a typical background arsenic concentration of 5 mg kg^{-1} in soil, and perhaps in most cases, inconsequential.

Another source of fertilizer arsenic to paddy fields is organic manures. In a Bengali context, while cattle manure can be high in arsenic as the animals are fed on rice straw which is elevated in arsenic, most cattle manure is dried and burnt as fuel (Pal et al. 2009), with there being a predisposition by farmers to use chemical (urea) as a nitrogen source and triple superphosphate as a phosphorus source.

Perhaps more significant with respect to arsenic input from manures is the use of municipal solid waste (MSW) where household waste is mixed with urban run-off and industrial effluent. Bhattacharyya et al. (2003) trialled MSW application (at a rate equivalent to adding 60 kg ha^{-1} nitrogen) to rice paddies in Calcutta, India. The MSW had 45 mg kg^{-1} arsenic, compared to a basal soil arsenic level of 19 mg kg^{-1}, with a cow dung manure treatment that was also included in the experiment having arsenic levels of 13 mg kg^{-1}. MSW consistently, over 3 years of trials, raised grain arsenic concentrations fivefold from 0.1 to 0.55 mg kg^{-1}, while cow dung resulted in a threefold increase to 0.35 mg kg^{-1}. While the manuring may have resulted in increased arsenic burdens, the arsenic application per plot would be relatively low, as for the inorganic phosphorus example outlined above in this section. It is more likely that the organic matter addition was responsible for this large increase in grain arsenic on MSW application as the same result also occurred with cow dung addition. Work by Williams et al. (2011) showed that soil organic matter (OM) content regulated arsenic mobilization in paddies, with increasing OM increasing mobilization and arsenic assimilation into grain. Mestrot et al. (2009, 2011) showed that direct manuring of paddy soil with cow dung greatly increased the mobilization of arsenic into soil pore water. This effect is likely due to increased microbial activity through the addition of easily degradable organic matter, which in turn leads to a more rapid development of anaerobic conditions and thus arsenic mobilization in flooded paddy soils (see Chap. 5). Disentangling the role of OM addition *per se*, rather than the additional arsenic inputs such addition brings, must be born into account when considering organic fertilization of paddies. These results indicate that fertilizing paddies with OM, even if it is low in arsenic, may result in elevated grain arsenic.

The use of the organo-arsenical Roxarsone (4-hydroxy-3-nitrobenzenearsonic acid) in poultry meat production has raised some concerns with respect to rice cultivation as manure emanating from production units where this practise is utilized are suspected to be applied to paddy fields (Liu et al. 2009; Wang et al. 2006). Although no direct evidence of paddy field contamination by Roxarsone has been identified, Wang et al. (2006) and Liu et al. (2009) have shown theoretically, through experimentation, that arsenic derived from Roxarsone can be assimilated by rice.

4.2.4 Point Source Industrial Pollution

Arsenic minerals are often associated with base metals (copper, lead, tin, tungsten and zinc) and precious ores (gold and silver) and, thus, the mining and processing of such ores can, and often does, lead to wide scale industrial pollution of soils, primarily with inorganic arsenic. More localized contamination of arsenic from industrial sources arises from the specific manufacture of arsenic products. These products may be inorganic or organic in nature.

4.2.4.1 Mining and Mineral Processing

The mining and processing, particularly smelting, of base and precious minerals have left a global pollution legacy on a massive scale. The sources of such mining derived pollution include: discarded mine spoil, runoff from mine tailing dams, the collapse of mining tailing dams with subsequent contaminated sediment pollution, and atmospheric deposition of arsenic resulting from smelting activity.

Perhaps the most catastrophic impact on agricultural land is the collapse of mine tailing dams. One such incident is reported by Liu et al. (2005) where, in 1985, a lead/zinc mine tailing dam in Chenzhou, Hunan, China collapsed under heavy rain contaminating a 0.4 km strip on either side of the downstream valley over a 15 km stretch with a 15 cm think layer of tailings sludge. Being a subsistence farming region the land was subsequently cultivated after mechanical removal of the bulk of the sludge. Agronomic soils in the impacted area after clean-up contained between 140 to 1,230 mg kg^{-1} arsenic. This compared to a reference non-impacted zone where paddy soil arsenic levels ranged from 80 to 120 mg kg^{-1}. Note that the reference paddy soils still have high arsenic concentration compared to "uncontaminated" global averages. This may be due to naturally geogenically elevated arsenic in soils due to being present in a mineralized zone, as well as due to diffuse pollution emanating from mining activity. Total arsenic concentration of rice grain grown in this contaminated zone averaged 0.93 mg kg^{-1}, which is very high compared to non-impacted rice from other parts of China. Zhu et al. (2008)

surveyed Chinese rice from non-mine and mine impacted regions of China and found that mine impacted rice had a higher total arsenic concentration, with a high percentage of this arsenic being inorganic.

Paddy soil and rice were investigated around an abandoned gold and silver mine in Myungbong, Korea, where surrounding paddies had become contaminated due to wind blown and mine spoil dust and due to contaminated water runoff from the spoil (Lee et al. 2006). Paddy soil (n = 5) had an average total arsenic content of 70 mg kg^{-1}, compared to a "natural" level in Korea of 10 mg kg^{-1}, with rice grain having an average total arsenic concentration of 0.41 mg kg^{-1} from the impacted soil, which is >4-fold higher than the baseline for Korea. Small scale manual mining for tin and tungsten in Dai Tu district, North Vietnam resulted in surface soil arsenic levels averaging 50 mg kg^{-1} (Ngoc et al. 2009), above what can be considered as the "uncontaminated" global soil background (see Sect. 4.2.1). No background reference soils were reported for this study, and neither were arsenic concentration in rice grain. Ruangwises and Saipan (2010) report that a large agronomic area (500 km^2) of Ron Phibun District, Thailand was contaminated with arsenic from tin mines abandoned for >100 years. Although they did not explicitly state soil or rice arsenic concentration, they measured inorganic arsenic concentration of the diet and found that dietary arsenic was highly elevated, and was the main source of inorganic arsenic intake.

Agricultural areas in around cities that are centres for mining and smelting can become elevated by arsenic with subsequent contamination of agricultural foodstuffs, including rice. An example of such peri-urban impacted sites is Chenzhou City, China where arsenic concentrations in agricultural soils have been recorded at 1,200 mg kg^{-1} (Liao et al. 2005).

4.2.4.2 Manufacture of Arsenicals

Although less well investigated, arsenicals can be released from factory sites producing arsenic based compounds. Such materials may have been historically dumped at the factory site or elsewhere. A number of infamous examples of such point source industrial contamination have arisen in Japan at Nakajo (Nakadaira et al. 1992) and Kizaki (Arao et al. 2009; Baba et al. 2008), with both instances based on organo-arsenical chemical weapons dump sites. Diphenylarsenic acid, phenylarsonic acid, methylphenylarsinic acid, dimethylphenylarsine oxide and methyldiphenylarsine oxide have found their way into groundwater used for agricultural irrigation, resulting in contamination of paddy soil and then, subsequently, rice grain.

4.2.5 Arsenical Pesticides

Rice is widely grown in what used to be the cotton belt of the USA (Williams et al. 2007a). Cotton is obviously not produced for human consumption, but rice is. It has been deemed acceptable to use arsenical based pesticides and defoliants on cotton, but this ignores crop rotation. There is a large literature on MMA and DMA impacts on rice production, where production is conducted on fields that have previously been used for cotton. In particular, there has been much research into selecting rice cultivars that are tolerant to MMA and DMA as these compounds lead to an induction of "straighthead disease" and decreased yield (see Chap. 6). Straighthead resistant cultivars have been selected to maintain yield in regions of the US where arsenical pesticides have led to large grain yield reductions (Williams et al. 2007). However, besides one paper on spray drift contamination (Wauchope et al. 1982), the consequences of previous arsenical pesticide application on grain arsenic has only recently been addressed. Williams et al. (2007) raised this issue by surveying the South Central US rice/cotton belt for rice grain arsenic and comparing the findings with the Northern Californian rice belt which suffered little from cotton production. Grain from the South Central rice belt had a mean of 0.30 mg kg^{-1} arsenic compared to Californian rice at 0.17 mg kg^{-1}. These findings were further supported by Zavala and Duxbury (2008). An average grain total arsenic level of 0.30 mg kg^{-1} is high in a global context.

It is only recently that direct evidence, rather than inference through survey (Williams et al. 2007a; Zavala and Duxbury 2008), has been provided to show that supplying rice with MMA under field conditions in the South Central US rice zone leads to elevated grain arsenic. Hua et al. (2011) found that field application of MMA to give a soil concentration of 20 mg kg^{-1}, under paddy (flooded) management resulted in grain total arsenic ranging from 0.45 to 1.5 mg kg^{-1} for three different cultivars, whereas grain arsenic was ~0.45 mg kg^{-1} for all cultivars when no MMA was applied.

Historically it is difficult to determine the extent of arsenical pesticide use on US rice grain as MMA is biodegradable to inorganic arsenic (see Chap. 5), and build up of MMA and inorganic arsenic after use will be dependent on soil type, field management and length and degree of application. Today, it is still not clear whether the high levels of arsenic in South Central US rice detected by Williams et al. (2007a) and Zavala and Duxbury (2008) is due to organo-arsenical use, or simply to inherent soil, management, genetic or climatological factors. Organo-arsenicals are now banned from use in the US (http://www.epa.gov/oppsrrd1/reregistration/organic_arsenicals_fs.html). The legacy of this application may take some time to resolve.

4.2.6 Contamination of Irrigation Water

As part of the Green Revolution groundwaters started to be widely used to irrigate rice during the dry season (Boro) in the Bay of Bengal, enabling this region to be largely self sufficient in rice production, with Boro season cultivation now taking place on ~50% of the land (Baffes and Gautam 2001). Boro rice varieties are improved from Aus landraces and yield better than wet season (Aman) post monsoonal cultivation due to greater light intensities, better regulation of water levels and absence of floodwater damage, amongst other factors. However, it is now being realized that Boro rice production poses major threats to the sustainable rice agriculture in the Bengal Delta through concerns regarding water use (Barr and Gowing 1998), nutrient supply and build-up of toxic elements, particularly arsenic (Neumann et al. 2010; Nobi and Dasgupta 1997; Williams et al. 2006). Around 50% of Bangladeshi groundwaters used for irrigation are elevated in arsenic leading to yield reduction and elevated grain arsenic (Williams et al. 2006), and extensive groundwater pumping is implicated in arsenic mobilization and further contamination of deep aquifers (Neumann et al. 2010); see Sect. 4.3 for a coverage of this topic.

4.2.6.1 Background to Arsenic Groundwaters in SE Asia

Arsenic groundwater chemistry depends on the geology of the aquifer and redox conditions. At pHs below 8, arsenate is absorbed to iron oxyhydroxide (e.g. FeOOH) surfaces (Smedley and Kinniburgh 2002). Above pH 10, arsenate exists almost entirely in solution phase. Arsenite also has a strong affinity to FeOOH, but a maximum of only ~80% is sorbed to FeOOH surfaces, and this falls dramatically at pHs above 9. Under reduced conditions, arsenite dominates. Arsenite has less affinity for iron oxyhydroxide phases, and hence is more mobile in aquifer pore waters. If arsenic is enriched in aquifer materials, it generally is only mobilised under reducing conditions where arsenic is mobilised as arsenite, or where pH is high, both arsenite and arsenate are mobile.

There are three major types of natural geological conditions giving rise to arsenic elevation of groundwaters:

- aquifers composed of rocks or sediments enriched with arsenic containing minerals of geogenic origin such as sulphide mineralization;
- aquifers containing sediments coated with iron oxyhydroxides (FeOOH) phases enriched in arsenic through hydrological action, with arsenic mobilised into porewater by reducing conditions;
- aquifers enriched in arsenic through high rates of evaporation in arid areas, leading to concentration of minerals in groundwaters; arsenic is mobile in such aquifers due to the high pH caused by concentration of alkali and alkali earth metals in solution.

In a number of SE Asia delta environments, deep fluvial and deltaic Pleitocene-Holocene sediments have accumulated (up to 10 km thick in Bangladesh) (Nickson et al. 2000). Figure 4.2 shows the extent of arsenic contamination of tubewell aquifers in Bangladesh. During glaciation, the river levels were 100 m lower than in interglacial times, and at this time of low sea-level, the sediments were flushed and oxidised, leading to iron oxyhydroxide precipitation on sediment surfaces. These sedimentary iron oxyhydroxides scavenge arsenic, with arsenic levels reaching up to 517 mg kg^{-1} in FeOOH phases (Nickson et al. 2000). Under reducing conditions, driven by microbial metabolism of sedimentary organic matter (present at up to 6% as C), where sulphate levels are low, insoluble Fe(III) is converted to soluble Fe(II), leading to the mobilisation of arsenic from the dissolved FeOOH phase.

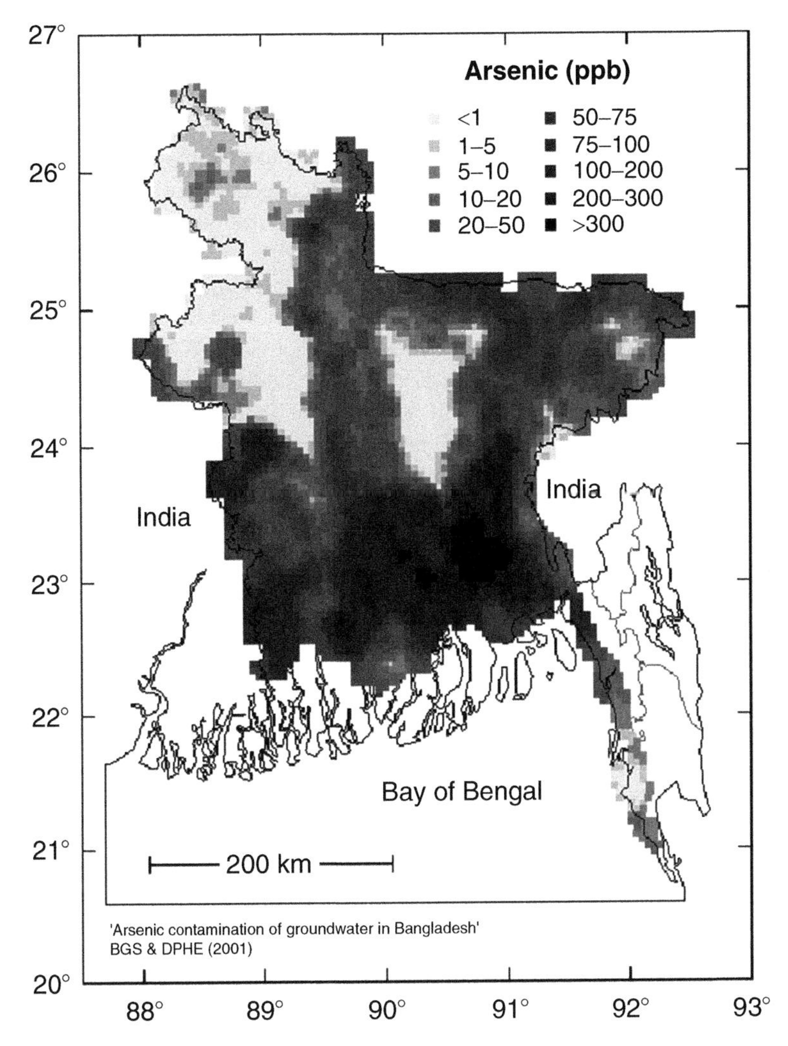

Fig. 4.2 The British Geological Survey's arsenic concentration in the groundwater of Bangladesh (Reproduced with permission)

Although traces of arsenic rich pyrites are found in the sediments, these are present at too small quantities for pyrite oxidation to contribute significantly to arsenic in groundwaters.

Similar geochemical conditions to Bangladesh alluvial sediments exist in the Red River alluvial tract in the city of Hanoi, Vietnam, where FeOOH reduction is thought to have led to the elevated arsenic levels recorded in groundwaters (Berg et al. 2008), as well as for the Mekong delta in Cambodia (Buschmann et al. 2008). Smedley and Kinniburgh (2002) outline that the reducing conditions observed in Bangladesh/West Bengal and Vietnam aquifers are similar to the regions of Taiwan, Northern China and Hungary which suffer from elevated arsenic in groundwaters.

4.2.6.2 Arsenic Build-up in Tubewell Irrigated Bengal Delta Soils

Early studies showed that arsenic was potentially building up in Bengal Delta soils irrigated with arsenic contaminated waters (Alam and Sattar 2000; Meharg and Rahman 2003; Ullah 1998), as arsenic in soil was elevated in groundwater-affected areas (Fig. 4.3). Meharg and Rahman (2003) calculated yearly loadings of arsenic into paddy soils based on tubewell water As concentrations, using a conservative estimate that 1,000 mm of tubewell water is used to irrigate rice fields per annum. If that water contains 0.1 mg As L^{-1} (equivalent to approximately 1 kg As per ha per annum), which is typical of many areas in Bangladesh-West Bengal, it was predicted that soil As concentrations should rise by 1 mg kg^{-1} per year. While a range of studies have found arsenic build-up in soils at particular study sites (Dittmar et al. 2010; Norra et al. 2005; Stroud et al. 2011; Van Geen et al. 2006), soil arsenic levels in Bangladesh paddies rarely exceed 50 mg kg^{-1} (Fig. 4.3), and the prediction of year after year arsenic build-up on soil proved more complex. Lu et al. (2009) found that baseline levels of arsenic varied over Bangladesh depending on whether sediments had been deposited into Holocene or Pleistocene contexts, suggesting that many of the soils of Bangladesh were naturally elevated in arsenic, similar to what has been the aquifer sediments (Smedley and Kinniburgh 2002). However, when paddy soil arsenic build-up over successive years was examined at a range of 20 high arsenic tube-well sites, at most sites there was not a year on year accumulation. This is illustrated by Fig. 4.4 which plots these impacted sites both against the period that the field had been under tubewell water irrigation and the arsenic levels in that tubewell water. It can be seen that the length in years under which a field has been used for Boro rice cultivation has little impact on arsenic accumulation, thus the correlation observed between arsenic levels in the paddy field and tubwell water arsenic is due to that current seasons irrigation. Thus, higher arsenic levels of rice found in the arsenic affected regions of Bangladesh compared to non-contaminated regions (Lu et al. 2009; Williams et al. 2006; Zavala et al. 2008) are primarily due to higher baseline soil arsenic levels and to current season irrigation water, though at some sites arsenic does build-up in the soils, contributing further to grain arsenic.

Mass balance studies by Dittmar et al. (2007) and Roberts et al. (2010) into the dynamics of arsenic in paddy fields irrigated with arsenic impacted tubewell waters during the following monsoon Aman production season found that much of the arsenic absorbed to the soil during Boro irrigation partitioned into the monsoonal floodwaters, being lost to the soil. Roberts et al. (2010) estimated that 13–62% of irrigation water arsenic added to soil during the Boro season could be lost to the following monsoonal floodwaters. Such loss of arsenic from paddy fields on a seasonal basis has been observed by Saha and Ali (2007) and Lu et al. (2009).

Besides losses to surface floodwaters, arsenic added in irrigation water may be leached to depth within the paddy profile. Dittmar et al. (2007) and Roberts et al. (2010) found that arsenic did not leach to depths greater than 30 cm at one site. A further loss mechanism of arsenic pumped onto paddy fields are flow-paths through field bunds (raised boundaries around fields) (Neumann et al. 2011). Actual loss of arsenic through bunds was not quantified on a mass balance basis, so it is difficult

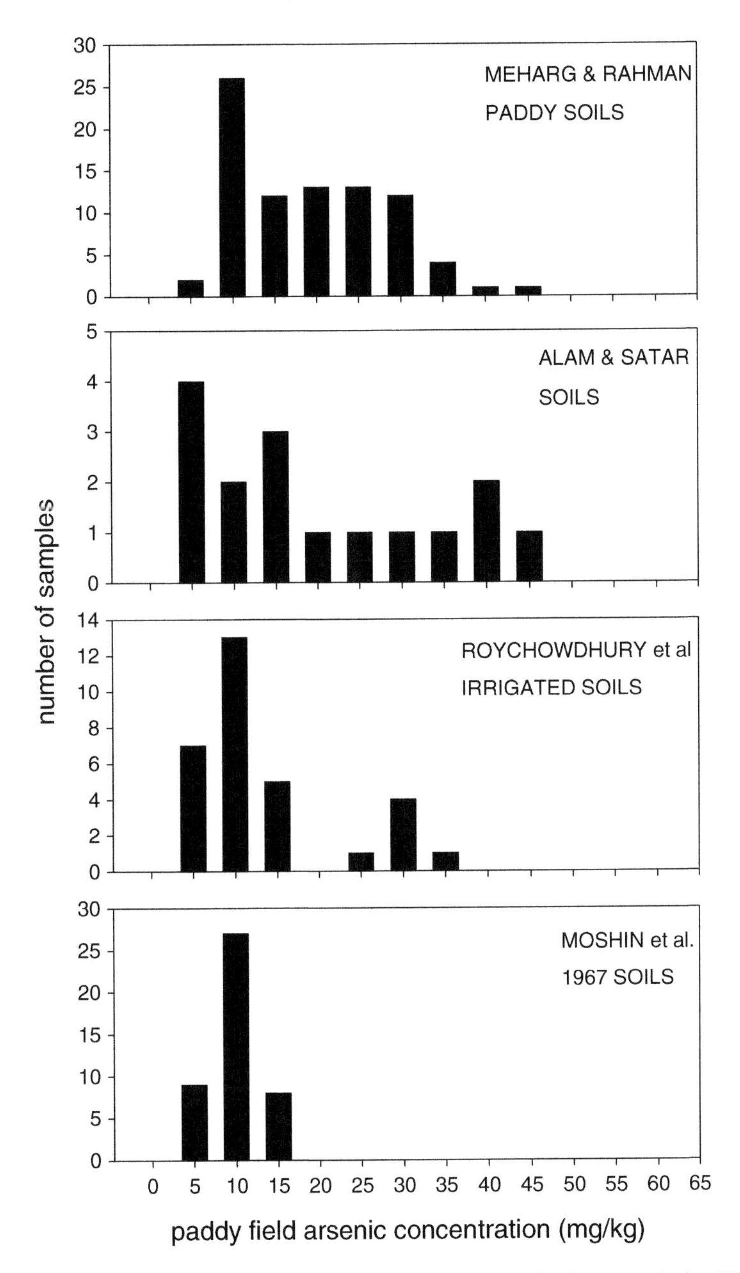

Fig. 4.3 Arsenic concentrations in Bangladesh and West Bengal soils (Data are obtained from the surveys of Alam and Sattar (2000), Meharg and Rahman (2003), Roychowdhury et al. (2003), with the data from Moshin et al. obtained from Ali et al. (2003))

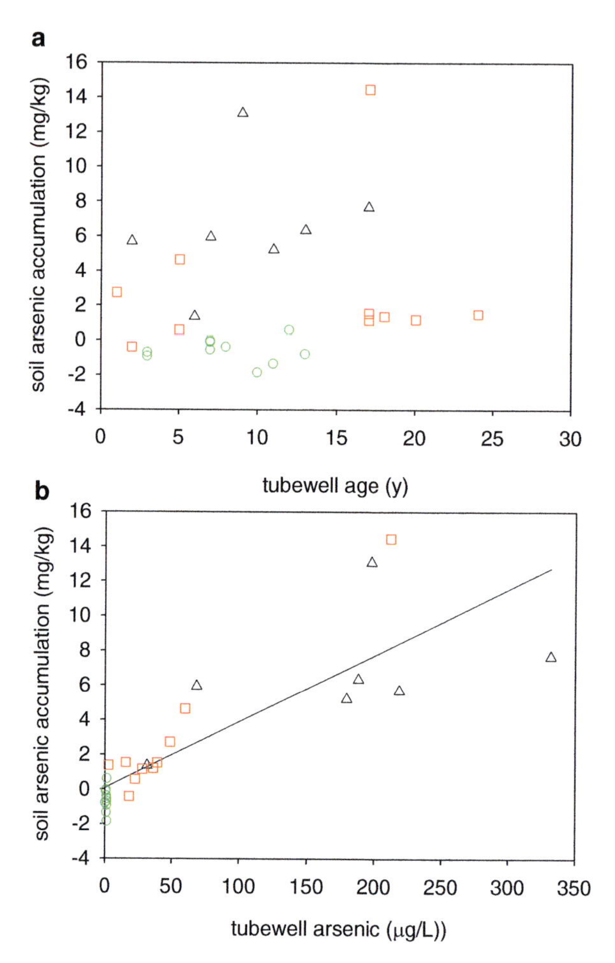

Fig. 4.4 Bangladesh paddy soil arsenic accumulation (irrigated – baseline control) plotted against tubewell age (**a**) and tubewell water arsenic concentration (**b**) from Lu et al. (2009). For the irrigated – baseline arsenic calculation for soil, year 1 and year 2 Gazipur (*circles*) irrigated field samples were first averaged. For Faridpur (*squares*), where two successive rice harvests were sampled, both years were averaged, but where the crop was under wheat for the second year, only year 1 data was used. For Jessore (*triangles*), year 2 data was used

to estimate overall contribution of bund-flow to arsenic mass-balances. Mestrot et al. (2011) showed that arsenic volatilization as arsines was not a major loss mechanisms of arsenic from Bengali paddies.

Lu et al. (2009) found a limited number of sites showing arsenic build up, but only 2 out of 20 impacted sites studied. At the sites where arsenic did build-up, there was a clear gradient of high arsenic at the point of entry of tubewell water into the paddy field, which declined with distance from this water supply, indicating that the tubewell water was the arsenic source. Others have found such trends in arsenic gradients building up near the site of tubewell entry (Dittmar et al. 2007; Roberts et al. 2010; Roychowdhury et al. 2003; Stroud et al. 2011). At some sites arsenic builds-up in soils (Dittmar et al. 2007), at others it does not (Lu et al. 2009). This difference in behaviour between sites may be due to edaphic factors or to the extent of monsoon flooding at each individual site. For example both Williams et al. (2011) and Stroud et al. (2011) found that the arsenic binding and release characteristics of Bangladeshi paddy soils varied greatly, which could relate to the arsenic retention capabilities of these soils. Understanding what factors affect arsenic build-up in paddy fields, or not, is essential if site management is to be instigated to negate this phenomenon.

4.3 Paddy Field Arsenic Cycling in a Wider Hydrological Context

There has been some debate as to the role of agricultural pumping of groundwater for Boro rice irrigation on wider arsenic dynamics within Bengali aquifers (Burgess et al. 2010; Neumann et al. 2010, 2011). Burgess et al. (2010) warn that switching from shallow groundwater (<200 m) to deep groundwater (>200 m), because deeper groundwater is generally lower in arsenic, is not a safe option as this will result in draw-down of arsenic from the shallow aquifer contaminating the deeper aquifer. Fendorf et al. (2010) also warn that deep aquifers should not be compromised by abstraction for irrigation, but should be preserved for drinking.

Paddy irrigation can lead to aquifer recharge as water pumped to the surface can flow through field bunds, as these are relatively freely draining, and back to the aquifer (Neumann et al. 2011). Neumann et al. (2011) have calculated that only 2% of the arsenic pumped to the surface returns to the aquifer through this field bund route, primarily as the arsenic is readily absorbed to bund sediment.

From an intensive study at one site, Neumann et al. (2010) looked at the interaction between pond waters (which they considered to be sources of organic carbon to the aquifer below, suggesting that this organic carbon was responsible for the mobilization of arsenic within the aquifer) and tubewells for agricultural pumping. They suggested that agricultural pumping caused drawdown of pond derived organic carbon to depths where arsenic could be mobilized. They also noted that, at the study site, arsenic levels were lower in water obtained under paddy fields as compared to under ponds. The observations at this site with regards to pond-tubewell interactions have yet to be replicated at other sites.

Besides direct contamination of paddy fields during each Boro season and incremental arsenic build-up at some field sites, it is clear that the scale of agricultural pumping for rice cultivation during the Boro season has major implications for current arsenic dynamics with shallow tubewell water and potential contamination of deeper tubewell supplies. Along with water conservation issues, it is imperative that the role of paddy field irrigation on wider arsenic dynamics is investigated, and with the evidence that is present to date, serious effort should be employed to reduce water usage from tubewells during the Boro season.

References

Alam MD, Sattar MA (2000) Assessment of arsenic contamination in soils and waters in some areas of Bangladesh. Water Sci Tech 42:185

Ali M, Ishiga H, Wakatsuki T (2003) Influence of soil type and properties on distribution and changes in arsenic contents of different paddy soils in Bangladesh. Soil Sci Plant Nutr 49:111–123

Allard B (1995) Groundwater. In: Salbu B, Steinnes E (eds) Trace elements in natural waters. CRC, Boca Raton, pp 151–176

Andreae MO (1980) Arsenic in rain and the atmospheric mass balance of arsenic. J Geophys Res 85:4512–4518

Arao T, Maejima Y, Koji B (2009) Uptake of aromatic arsenicals from soil contaminated with diphenylarsinic acid by rice. Environ Sci Technol 43:1097–1101

Baba K, Arao T, Maejima Y, Watanabe E, Eun H, Ishizaka M (2008) Arsenic speciation in rice and soil containing related compounds of chemical warfare agents. Anal Chem 80:5768–5775

Baffes J, Gautam M (2001) Assessing the sustainability of rice production growth in Bangladesh. Food Policy 26:515–542

Ballantyne JM, Moore JN (1988) Arsenic geochemistry in geothermal systems. Geochem Cosmochim Acta 52:475–483

Barr JJF, Gowing JW (1998) Rice production in floodplains: issues for water management in Bangladesh. In: Pereira LS, Gowing JG (eds) Water and the environment: innovative issues in irrigation and drainage. E & FN Spon, London, pp 308–317

Berg M, Trang PPTK, Stengel C, Buschmann J, Viet PH, Van Dan N, Giger W, Stueben D (2008) Hydrological and sedimentary controls leading to arsenic contamination of groundwater in the Hanoi area, Vietnam: The impact of iron-arsenic ratios, peat, river bank deposits, and excessive groundwater abstraction. Chem Geol 249:91–112

Bhattacharyya P, Ghosh AK, Chakraborty A, Chakrabrti K, Tripathy S, Powell MA (2003) Arsenic uptake by rice and accumulation in soil amended with municipal solid waste compost. Commun Soil Sci Plant Anal 34:2779–2790

Burgess WG, Hoque MA, Michael HA, Voss CI, Breit GN, Ahmed KM (2010) Vulnerability of deep groundwater in the Bengal Aquifer System to contamination by arsenic. Nat Geosci 3:83–87

Buschmann J, Berg M, Stengel C, Winkel L, Sampson ML, Trang PTK, Viet PH (2008) Contamination of drinking water resources in the Mekong delta floodplains: arsenic and other trace metals pose derious health risks to population. Environ Int 34:756–764

Charter RA, Tabatabai MA, Schafer JW (1995) Arsenic, molybdenum, selenium and tungsten contents of fertilizers and phosphate rocks. Commun Soil Sci Plant Anal 26:3051–3062

Chen WP, Krage N, Wu L, Pan G, Khosrivafard M, Chang AC (2008) Arsenic, cadmium, and lead in California cropland soils: role of phosphate and micronutrient fertilizers. J Environ Qual 37:689–695

Clayton CA, Pellizzari ED, Whitmore RW, Perritt RL, Quackenboss JJ (1999) National Human Exposure Assessment Survey (NHEXAS): distributions and associations of lead, arsenic and volatile organic compounds in EPA region 5. J Expo Anal Env Epidemiol 9:381–392

Colbourn P, Alloway BJ, Thornton I (1975) Arsenic and heavy metals in soils associated with regional geochemical anomalies in south-west England. Sci Total Environ 4:359–363

Dittmar J, Voegelin A, Roberts LC, Hug SJ, Saha GC, Ali MA, Badruzzaman ABM, Kretzschmar R (2007) Spatial distribution and temporal variability of arsenic in irrigated rice fields in Bangladesh. 2. Paddy soil. Environ Sci Technol 41:5967–5972

Dittmar J, Voegelin A, Roberts LC, Hug SJ, Saha GC, Ali MA, Badruzzaman ABM, Kretzschmar R (2010) Arsenic accumulation in a paddy field in Bangladesh: seasonal dynamics over a three-year monitoring period. Environ Sci Technol 44:2925–2931

Fendorf S, Holly M, van Geen A (2010) Spatial and temporal variations of groundwater arsenic in South and Southeast Asia. Science 328:1123–1127

Ferre-Huguet N, Marti-Cid R, Schluhmacher M, Domingo JL (2008) Risk assessment of metals from consuming vegetables, fruits and rice grown on soils irrigated with waters of the Ebro River in Catalonia, Spain. Biol Trace Elem Res 123:66–79

Geiszinger A, Goessler W, Kosmus W (2002) Organoarsenic compounds in plants and soil on top of an ore vein. Appl Organomet Chem 16:245–249

Goldschmidt VM (1958) In: Muir A (ed) Geochemistry. Oxford University Press, Oxford, pp 468–475

Hale M (1981) Pathfinder applications of arsenic, antimony and bismuth in geochemical exploration. J Geochem Explor 15:307–327

Hua B, Yan WG, Wang JM, Deng BL, Yang J (2011) Arsenic accumulation in rice grains: effects of cultivars and water management practices. Environ Eng Sci 28:591–596

Hubert JS, Candelaria RM, Rosenblum B, Munoz R, Applegate HG (1981) A survey of ambient air levels of arsenic and cadmium in El Paso, Texas from 1972–1979. JAPCA J Air Poll Cont Assoc 31:261–262

Koljonen T (1992) Geochemical atlas of Finland. Geological Survey of Finland, Espoo

Lee HS, Cho YH, Park SO, Kye SH, Kim BH, Hahm TS, Kim M, Lee JO, Kim C (2006) Dietary exposure of the Korean population to arsenic, cadmium, lead and mercury. J Food Composit Anal 19:31–37

Liao X-Y, Chen T-B, Liu Y-R (2005) Soil As contamination and its risk assessment in areas near the industrial districts of Chenzhou City, Southern China. Environ Int 2005:791–798

Liu H, Probst A, Liao B (2005) Metal contamination of soils and crops affected by the Chenshou lead/zinc mine spill (Hunan, China). Sci Total Environ 339:153–166

Liu CW, Lin CC, Jang CS, Sheu GR, Tsui L (2009) Arsenic accumulation by rice grown in soil treated with roxarsone. J Plant Nutr Soil Sci 172:550–556

Lu Y, Adomako EE, Solaiman ARM, Islam RM, Deacon C, Williams PN, Rahman GK, Meharg AA (2009) Baseline soil variation is a major factor in arsenic accumulation in Bengal delta paddy rice. Environ Sci Technol 43:1724–1729

Macedo SM, de Jesus RM, Garcia KS, Hatje V, Queiroz AFD, Ferreira SLC (2009) Determination of total arsenic and arsenic (III) in phosphate fertilizers and phosphate rocks by HG-AAS after multivariate optimization based on Box-Behnken design. Tantala 80:974–979

Matschullat J (2000) Arsenic in the geosphere – a review. Sci Total Environ 249:297–312

Meharg AA, Rahman MM (2003) Arsenic contamination of Bangladesh paddy field soils: implications for rice contribution to arsenic consumption. Environ Sci Technol 37:229–234

Meharg AA, Williams PN, Adamako E, Lawgali YY, Deacon C, Villada A, Cambell RCJ, Sun GX, Zhu YG, Feldmann J, Raab A, Zhao FJ, Islam R, Hossain S, Yanai J (2009) Geographical variation in total and inorganic arsenic content of polished (white) rice. Environ Sci Technol 43:1612–1617

Mestrot A, Uroic K, Plantevin T, Islam Md R, Krupp E, Feldmann J, Meharg AA (2009) Quantitative and qualitative trapping of arsines deployed to assess loss of volatile arsenic from paddy soil. Environ Sci Technol 43:8270–8275

Mestrot A, Feldmann J, Krupp EM, Sumon MH, Roman-Ross G, Meharg AA (2011) Field fluxes and speciation of arsines emanating from soil. Environ Sci Technol 45:1798–1804

Nakadaira H, Yamamoto M, Katoh K (1992) Arsenic levels in soil of a town polluted 35 years ago (Nakajo, Japan). Bull Environ Contam Toxicol 55:650–657

Neumann RB, Ashfaque KN, Badruzzaman ABM, Ali MA, Shoemaker JK, Harvey CF (2010) Anthropogenic influences on groundwater arsenic concentrations in Bangladesh. Nat Geosci 3:46–52

Neumann RB, St Vincent AP, Roberts LC, Badruzzaman ABM, Ali MA, Harvey CF (2011) Rice field geochemistry and hydrology: an explanation for why groundwater irrigated fields in Bangladesh are net sinks of arsenic from groundwater. Environ Sci Technol 45:2072–2078

Ngoc KC, Nguyen NV, Dinh BN, Thanh SL, Tanaka S, Kang Y, Sakutai K, Iwasaki K (2009) Arsenic and heavy metal concentrations in agricultural soils around tin and tungsten mines in the Dai Tu district, N. Vietnam. Water Air Soil Pollut 197:75–89

Nickson RT, McArthur JM, Ravenscroft P, Burgess WG, Ahmed KM (2000) Mechanisms of arsenic release to groundwater, Bangladesh and West Bengal. Appl Geochem 15:403–413
Nobi N, Dasgupta A (1997) Simulation of regional flow and salinity intrusion in an integrated stream-aquifer system in coastal region: Southwest Region of Bangladesh. Ground Water 35:786–796
Nordstrom DK (2002) Worldwide occurrences of arsenic in ground water. Science 296:2143–2145
Norra S, Berner ZA, Agarwala P, Wagner F, Chandrasekharam D, Stubben D (2005) Impact of irrigation with As rich groundwater on soil and crops: a geochemical case study in West Bengal delta plain, India. Appl Geochem 20:1890–1906
Norton GJ, Islam R, Duan G, Lei M, Zhu YG, Deacon CM, Moran AC, Islam S, Zhao FJ, Stroud JL, McGrath SP, Feldmann J, Price AH, Meharg AA (2010) Arsenic shoot - grain relationships in field grown rice cultivars. Environ Sci Technol 44:1471–1477
Pal A, Chowdhury UK, Mondal D, Das B, Nayak B, Ghosh A, Maity S, Chakraborti D (2009) Arsenic burden from cooked rice in the populations of arsenic affected and nonaffected areas and Kolkata City in West-Bengal, India. Environ Sci Technol 43:3349–3355
Roberts LC, Hug SJ, Dittmar J, Voegelin A, Kretzschmar R, Wehrli B, Cirpka OA, Saha GC, Ali MA, Badruzzaman ABM (2010) Arsenic release from paddy soils during monsoon flooding. Nat Geosci 3:53–59
Roychowdhury T, Tokunaga H, Ando M (2003) Survey of arsenic and other heavy metals in food composites and drinking water and estimation of dietary intake by the villagers from an arsenic-affected area of West Bengal, India. Sci Total Environ 308:15–35
Ruangwises S, Saipan P (2010) Dietary intake of total and inorganic arsenic by adults in an arsenic contaminated area of Ron Phibun District, Thailand. Bull Environ Contam Toxicol 84:274–277
Saha GC, Ali MA (2007) Dynamics of arsenic in agricultural soils irrigated with arsenic contaminated groundwater in Bangladesh. Sci Total Environ 379:180–189
Schoof RA, Yost LJ, Eickhoff J, Crecelius EA, Irgolic K, Goessler W, Guo HR, Greene H (1998) Dietary arsenic intake in Taiwanese districts with elevated arsenic in drinking water. Hum Ecol Risk Assess 4:117
Shotyk W, Cheburkin AK, Appleby PG, Fankhauser A, Kramers JD (1996) Two thousand years of atmospheric arsenic, antimony, and lead deposition recorded in an ombrotrophic peat bog profile, Jura Mountains, Switzerland. Earth Planet Let 145:1–4
Smedley PL, Kinniburgh DG (2002) A review of the source, behaviour and distribution of arsenic in natural waters. Appl Geochem 17:517–568
Stroud JL, Norton GJ, Islam MR, Dasgupta T, White RP, Price AH, Meharg AA, McGrath SP, Zhao FJ (2011) The dynamics of arsenic in four paddy fields in the Bengal delta. Environ Pollut 159:947–953
Ullah SM (1998) In: International conference on arsenic pollution of groundwater in Bangladesh: causes, effects and remedies. Dhaka Community Hospital, Dhaka, Bangladesh, 8–12 Feb 1998 (Abstracts), 133 pp
USDA (2006) Foreign Agricultural Service, Global Agriculture Information Network Report, Peoples Republic of China, FAIRS Product Specific Maximum Levels of Contaminants in Foods
van Geen A, Zheng Y, Cheng Z, He Y, Dhar RK, Garnier JM, Rose J, Seddique A, Hoque MA, Ahmed KM (2006) Impact of irrigating rice paddies with groundwater containing arsenic in Bangladesh. Sci Total Environ 367:769–777
Wang FM, Chen ZL, Zhang L, Gao YL, Sun YX (2006) Arsenic uptake and accumulation in rice (*Oryza sativa* L.) at different growth stages following soil incorporation of roxarsone and arsanilic acid. Plant Soil 285:359–367
Wauchope RD, Richard EP, Hurst HR (1982) Effects of simulated MSMA drift on rice (Oryza-sativa). 2. Arsenic residues in foliage and grain and relationships between arsenic residues, rice toxicity symptoms, and yields. Weed Sci 30:405–410

Williams PN, Islam MR, Adomako EE, Raab A, Hossain SA et al (2006) Increase in rice grain arsenic for regions of Bangladesh irrigating paddies with elevated arsenic in groundwaters. Environ Sci Technol 40:4903–4908
Williams PN, Raab A, Feldmann J, Meharg AA (2007a) Market basket survey shows elevated levels of as in South Central US processed rice compared to California: consequences for human dietary exposure. Environ Sci Technol 41:2178–2183
Williams PN, Villada A, Deacon C, Raab A, Figuerola J et al (2007b) Greatly enhanced arsenic shoot assimilation in rice leads to elevated grain levels compared to wheat and barley. Environ Sci Technol 41:6854–6859
Williams PN, Lei M, Sun G-X, Huang Q, Lu Y, Deacon C, Meharg AA, Zhu Y-G (2009) Occurrence and partitioning of cadmium, arsenic and lead in mine impacted paddy rice: Hunan, China. Environ Sci Technol 43:637–642
Williams PN, Zhang H, Davison W, Meharg AA, Sumon MH, Norton G, Brammer H, Islam R (2011) Organic matter – solid phase interactions are critical for predicting arsenic release and plant uptake in Bangladesh paddy soils. Environ Sci Technol 45:6080–6087
Winkel L, Berg M, Amini M, Hug SJ, Johnson CA (2008) Predicting groundwater arsenic contamination in Southeast Asia from surface parameters. Nat Geosci 1:536–542
World Health Organization (2004) IARC, Working Group on some drinking water disinfectants and contaminants, including arsenic, vol 84. Lyon. Monograph 1
Zavala YJ, Duxbury JM (2008) Arsenic in rice: I. Estimating normal levels of total arsenic in rice grain. Environ Sci Technol 42:3856–3860
Zhu Y-G, Sun G-X, Lei M, Teng M, Liu Y-X, Chen N-C, Hong W-L, Carey AM, Meharg AA, Williams PN (2008) High percentage inorganic arsenic content of mining impacted and nonimpacted Chinese rice. Environ Sci Technol 42:5008–5013

Chapter 5
Biogeochemistry of Arsenic in Paddy Environments

5.1 Overview of the Biogeochemistry of Paddy Soil

Flooding is a common practice for the cultivation of paddy rice. Rice can be grown aerobically, but yield tends to be smaller than flooded rice possibly due to the build-up of pathogens and nematodes and generally lower bioavailability of nutrients (e.g. phosphorus) under aerobic conditions (Ventura et al. 1981). Flooding soil has a profound impact on the biogeochemical cycles of major and trace elements primarily through influencing the reduction-oxidation (redox) reactions. For a comprehensive discussion on this topic, readers are referred to the book by Kirk (2004). A brief overview is presented in this section.

Flooding soil causes a rapid decrease in the redox potential due to depletion of various electron acceptors, or oxidants. Box 5.1 describes briefly the relationships between the electron activity p*e*, the equilibrium constant *K* of redox half-reaction and the redox potential Eh. The range of p*e* in soils typically varies from +13 to −6, which can be divided into three parts corresponding to oxic soils (p*e*>7 at pH 7), suboxic soil (+2<p*e*<+7 at pH 7) and anoxic soils (p*e*<+2 at pH 7) (Sposito 1989) (see also Fig. 5.1). As p*e* decreases (the electron activity increases), a sequence of reduction half-reactions (also called redox couples) proceed as predicted by thermodynamics (Fig. 5.1). These are called half-reactions because two of them must be coupled to complete an overall redox reaction; in flooded soils oxidation of organic matter often provides the other half-reaction.

Upon flooding, O_2 in soil is depleted quickly, becoming undetectable usually within a day except in the top layer of around 10 mm depth (Kirk 2004). This is followed by reduction of nitrate, which is exhausted within days. After nitrate, manganese oxides and then iron oxides/hydroxides are reduced. Because different soils may contain different forms of manganese and iron oxides/hydroxides, which also vary in the degree of crystallinity (e.g. amorphous, meta-stable poorly crystalline or crystalline), the p*e* range within which they are reduced may differ from soil to soil. Reduction of ferric Fe oxides/hydroxides generally indicates the development of anoxic conditions. After the reduction of ferric Fe, sulphate is reduced to

A.A. Meharg and F.J. Zhao, *Arsenic & Rice*, DOI 10.1007/978-94-007-2947-6_5,

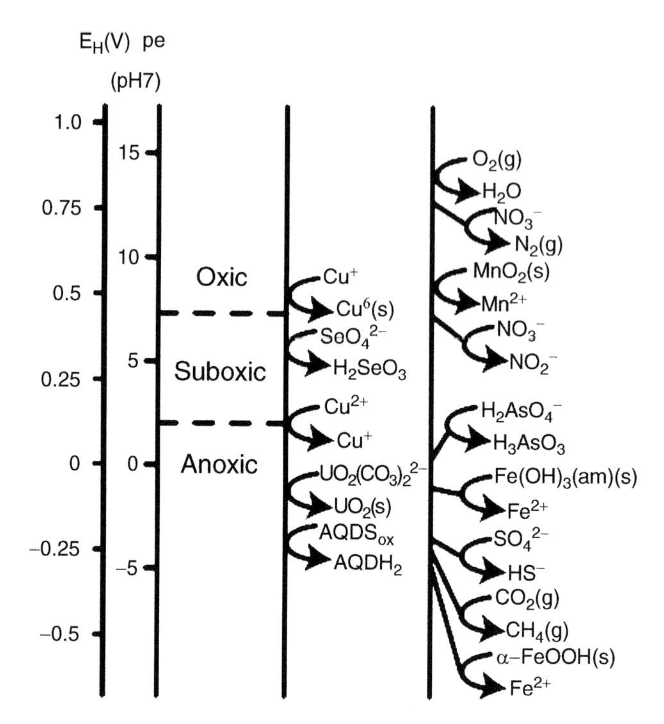

Fig. 5.1 Redox ladder showing examples of environmentally relevant redox couples. Couples were calculated at pH 7 using standard techniques. Concentrations of all dissolved constituents in each half reaction are 1 M except for Fe^{2+} (1×10^{-5} M), U^{6+} (1×10^{-6} M), and CO_3^{2-} (3×10^{-3} M). $AQDS_{ox}$ and $AQDH_2$ are oxidized and reduced forms of hydroquinone moieties, respectively (Reprinted with permission from Borch et al. (2010). Copyright (2010) American Chemical Society)

sulphide and with further decrease in p*e*, methane (CH_4) is produced. During each reduction step, p*e* is poised within the range of that reaction until the oxidant of the half-reaction is exhausted. Figure 5.1 also shows some reduction half-reactions of redox active trace elements, including copper, selenium, uranium and arsenic. Reactions involving arsenic are discussed in Sect. 5.2.

It should be noted that many of the redox reactions in soil are catalysed by appropriate microorganisms; without their involvement the reactions would proceed very slowly (Kirk 2004). Microbes catalyse the reaction by mediating electron transfer and, therefore, greatly increase the rate of reaction without changing its equilibrium constant. Some reactions can be catalysed by many different microorganisms, e.g. decomposition (oxidation) of organic matter, whereas other reactions require specific classes of microbes (e.g. denitrification, reduction of ferric iron, methane production). These specific microorganisms become active when p*e* falls within the range of the reactions they catalyse.

Another important point to remember is that nearly all reduction half-reactions consume protons (see Eq. 5.1 in Box 5.1). This explains why pH in acidic soils

Box 5.1 Redox Reactions: Relationships Between log K, pe, ΔG_r^o and Eh (Based on Bartlett and James 1993; Sparks 2003)

Most of reduction half-reactions can be written as the flowing generic formula:

$$a(\mathrm{Ox}) + b(e^-) + c(\mathrm{H}^+) = d(\mathrm{Red}) + f(\mathrm{H_2O}) \quad (5.1)$$

where a, b, c, d and f are reaction coefficients for the oxidised species (Ox), electron (e^-), proton (H^+), reduced species (Red) and water (H_2O), respectively. The equilibrium constant K is:

$$K = (\mathrm{Red})^d(\mathrm{H_2O})^f / (\mathrm{Ox})^a(e^-)^b(\mathrm{H}^+)^c \quad (5.2)$$

Applying log transformation and (H_2O) = 1, Eq. 5.2 becomes

$$\log K = d\log(\mathrm{Red}) - a\log(\mathrm{Ox}) - b\log(e^-) - c\log(\mathrm{H}^+) \quad (5.3)$$

$$\log K = d\log(\mathrm{Red}) - a\log(\mathrm{Ox}) + b(pe) + c(\mathrm{pH}) \quad (5.4)$$

$$pe = [\log K - d\log(\mathrm{Red}) + a\log(\mathrm{Ox})] / b - c / b(\mathrm{pH}) \quad (5.5)$$

where pe and pH are – log of electron activity and proton activity, respectively. Equation 5.5 allows one to construct the pe – pH diagram for specific redox half-reaction couples. For a half-reaction involving one electron transfer and one proton consumption, and when (Red) = (Ox) and $d = a$, Eq. 5.5 can be simplified as

$$pe + \mathrm{pH} = \log K \quad (5.6)$$

and at pH = 0 (1 M H^+ activity),

$$pe = \log K \quad (5.7)$$

The equilibrium constant K is related to the Gibbs free energy change (ΔG_r^o) for a given half-reaction as follows,

$$\Delta G_r^o = -RT \ln K = -5.71 \log K \quad (5.8)$$

where ΔG_r^o is the Gibbs free energy of reaction under standard conditions (T = 298.15 K and 100 kPa), R is the universal gas law constant (0.008314 kJ mol^{-1} K^{-1}), and ln (x) = 2.303 log (x). One can calculate log K from ΔG_r^o,

$$\Delta G_r^o = \sum \Delta G_{f\ \mathrm{products}}^o - \sum \Delta G_{f\ \mathrm{reactants}}^o \quad (5.9)$$

(continued)

Box 5.1 (continued)

ΔG_f^o is the free energy of formation (available from physical-chemical databases), and by convention, ΔG_f^o for H^+ and e^- are zero.
log K is also related to measured electrochemical potential, Eh, according to the following equation,

$$\Delta G_r^o = -nF \text{ Eh} \quad (5.10)$$

where n is the number of e^- transferred/mole and F the Faraday constant (9.65×10^4 C mol^{-1}). From Eqs. 5.8 and 5.10, at $T = 298$ K, and if $n = 1$,

$$\text{Eh (V)} = (-RT \ln K)/nF = 0.059 \log K \quad (5.11)$$

At the equal activity of (Red) and (Ox), Eq. 5.11 can be expressed as

$$\text{Eh (V)} = 0.059(pe + \text{pH}) \quad (5.12)$$

log K values for common reduction half-reactions in soil can be found in Bartlett and James (1993) and Sparks (2003). They indicate whether a particular half-reaction will occur under certain conditions based on the thermodynamic considerations only. Higher log K values indicate greater "ease of reduction" of an oxidant to its reduced form. However, a thermodynamically favoured reaction may occur very slowly due to kinetic hindrance or because reduction and oxidation half-reactions do not couple well to each other.

tends to increase to near neutrality after flooding (Kirk 2004). On the other hand, CO_2 produced from microbial respiration and fermentation escapes flooded soils very slowly. The accumulation of CO_2 lowers the pH of alkaline soils and curbs the increase in the pH of acidic soils (Kirk 2004). As a result, soil pH tends to converge in the range of 6.5–7 following flooding.

While the redox active elements are directly affected by redox reactions, there are also indirect effects that can influence the solubility and bioavailability of both redox active and non-active elements. For example, reduction of iron oxides/hydroxides can release the elements sorbed on the mineral surfaces. It is often observed that flooding soils increases phosphate bioavailability to rice plants mainly because of this effect (Kirk 2004; Larsen 1969). This is also an important mechanism that leads to increased arsenic bioavailability (Sect. 5.2). In contrast, bioavailability of some trace elements may decrease considerably after flooding due to the formation of carbonate or sulphide minerals (e.g. $ZnCO_3$, CdS).

5.2 Arsenic Transformations in Paddy Soil

Figure 5.2 shows the key processes of arsenic transformation in soil, some of which are enhanced under the flooded paddy environment, e.g. reduction, desorption, methylation and volatilisation, leading to profound changes in the arsenic bioavailability to plants. These processes are discussed in more details below.

5.2.1 *Arsenic Reduction and Oxidation*

5.2.1.1 Arsenic Mobilisation Induced by Soil Flooding

The log *K* value for the following arsenate reduction half-reaction is 16.5.

$$1/2\,AsO_4^{3-} + e^- + 2H^+ = 1/2\,AsO_2^- + H_2O$$

At equal concentrations (0.1 mM) of arsenate and arsenite, p*e* of the above reduction half-reaction would be 6.5 and 2.5 at pH 5 and 7, respectively (Bartlett and James 1993) (see also Box 5.1). The p*e* – pH diagram for the predominant aqueous species of arsenic at equilibrium is shown in Fig. 5.3. In this diagram, the dividing lines represent 50%/50% of the two neighbouring species; above the line the oxidised species dominates and *vice versa*. It is clear from thermodynamic considerations that arsenate is stable in the oxic environment whereas arsenite prevails in the suboxic and anoxic environments. The p*e* of the arsenate reduction places this

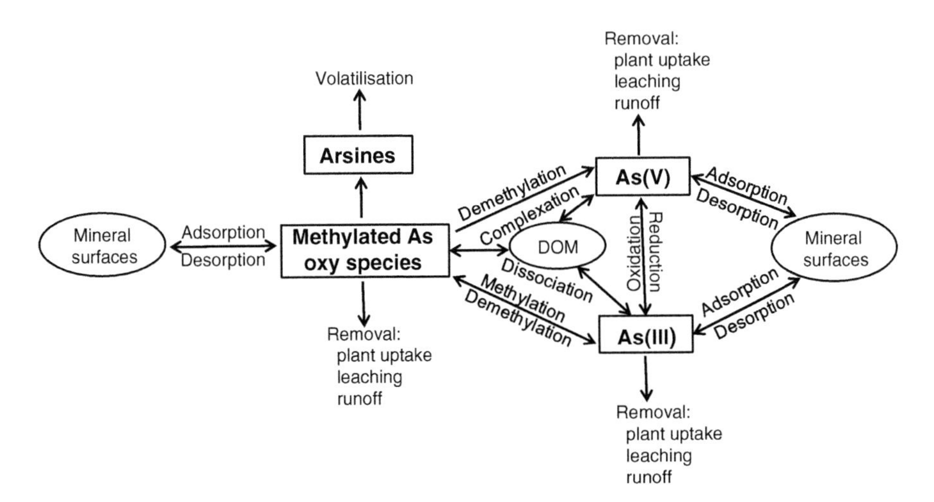

Fig. 5.2 Key processes of the arsenic cycle in soil

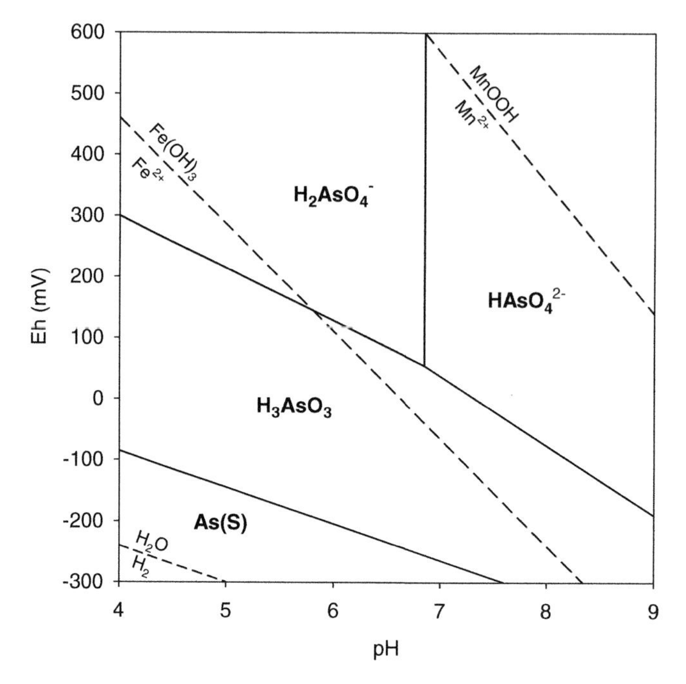

Fig. 5.3 Eh-pH diagram for the As-H_2O system. Activities of As, Mn, and Fe were all taken to be 10^{-4} M (Redrawn with permission from Masscheleyn et al. (1991). Copyright (1991) American Chemical Society)

half-reaction after the reduction of nitrate and manganese oxide, but within the similar p*e* range as the reduction of ferric hydroxide (depending on pH). Thermodynamic calculations by Kocar and Fendorf (2009) show that arsenate reduction is favourable over a wide range of field conditions and should proceed before Fe(III) and sulphate reduction.

The effects of redox potential and pH on arsenic mobilisation and speciation in soil solution have been investigated in a number of studies (e.g. Marin et al. 1993; Masscheleyn et al. 1991; Onken and Hossner 1996). Masscheleyn et al. (1991) incubated a soil heavily contaminated with arsenic from a cattle arsenic dipping vat, with Eh and pH in the soil-water suspension being controlled at different values. Soluble arsenite increased rapidly when Eh dropped to below 200 mV (~p*e* 3), whilst soluble arsenate decreased concurrently. In the Eh range of −200 to +100 mV, arsenite accounted for 87–98% of the total soluble As. Solubility of arsenate was generally low, except at high pH and Eh values (see Sect. 5.2.2 for more discussion). The overall effect was a dramatic mobilisation of arsenic, predominantly as arsenite, into the soil solution phase once suboxic/anoxic conditions developed in the soil. This effect has been demonstrated in other studies using either paddy or non-paddy soils under laboratory, greenhouse or field conditions (Arao et al. 2009; Li et al. 2009; Marin et al. 1993; Onken and Hossner 1996; Signes-Pastor et al.

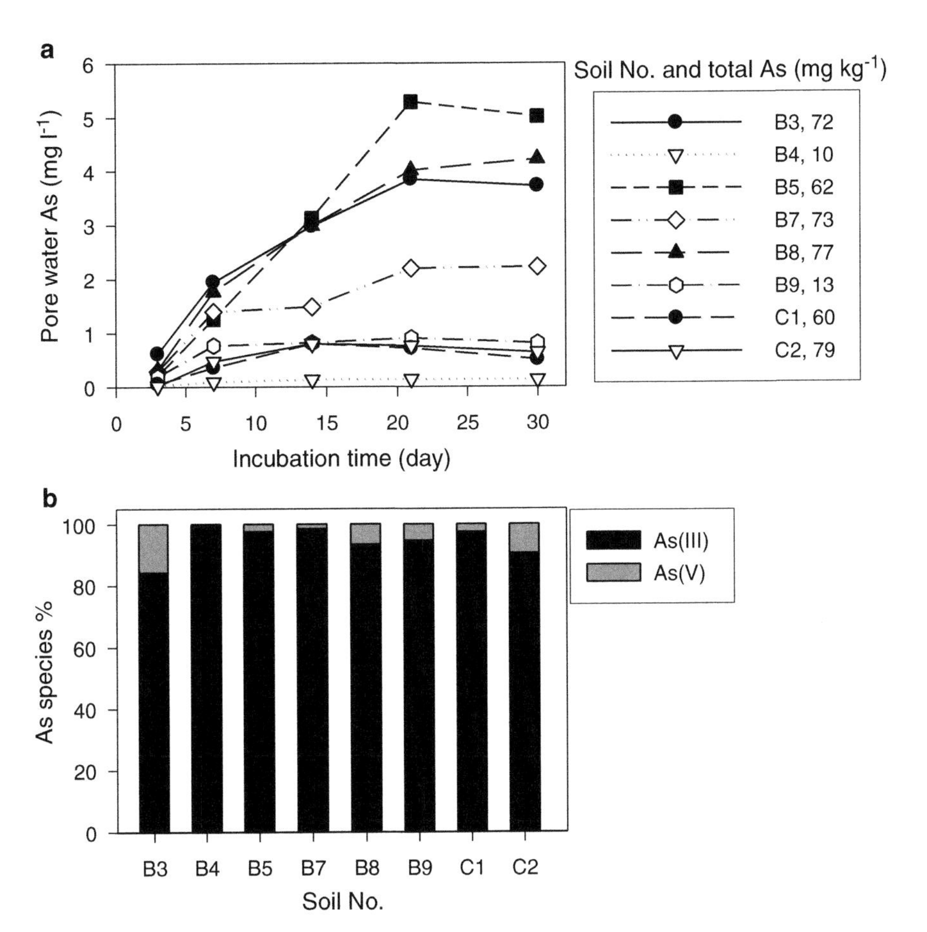

Fig. 5.4 Time-course of arsenic mobilisation into soil pore water after flooding in a laboratory incubation experiment (**a**) and arsenic species in pore water on day 21 (**b**). Soils *B3–B9* are from Bangladesh with the main source of arsenic contamination being irrigation of groundwater. Soil *C1–C2* are from China with the main source of arsenic contamination being mining (C1) or geogenic (C2) (Redrawn with permission from Stroud et al. (2011a). Copyright (2011) American Chemical Society)

2007; Stroud et al. 2011a; Takahashi et al. 2004; Xu et al. 2008; Yamaguchi et al. 2011).

In the example shown in Fig. 5.4, a number of As-contaminated paddy soils from Bangladesh and China were incubated under flooded conditions to monitor arsenic mobilisation into the soil pore water (Stroud et al. 2011a). By day 7 pore-water Eh had dropped to below 200 mV and the arsenic concentration in the pore water increased markedly reaching a plateau at day 21. Also, most (>80%) of the arsenic mobilised into the pore water was in the form of arsenite. The extent of arsenic mobilisation varied greatly among the soils. This variation can be partly explained by the total concentration and the lability of arsenic in the soils. Four of the Bangladeshi paddy soils that have been heavily contaminated (total arsenic 62–77 mg kg^{-1}) due to long-term irrigation of As-laden groundwater mobilised large concentrations of arsenic into the pore water (2–5 mg l^{-1}), whereas two Chinese paddy soils contaminated by mining or geogenic sources containing comparable levels of total arsenic (60–79 mg kg^{-1}) mobilised much lower levels of arsenic (0.5–0.8 mg l^{-1}). These results show that the Bangladeshi paddy soils contaminated by irrigation-water arsenic have a greater arsenic lability.

5.2.1.2 Mechanisms of Flooding-Induced Arsenic Mobilisation

There are two main reasons why arsenic solubility increases in response to the development of anaerobic conditions in flooded soil. Firstly, some of iron oxides/hydroxides are reduced and released into the solution phase, a process called reductive dissolution. Because these iron oxide minerals are an important host phase of arsenic in soil, their dissolution releases the sorbed arsenic into the solution phase. Cummings et al. (1999) demonstrated that the dissimilatory iron reducing bacterium *Shewanella alga* strain BrY was able to reduce Fe(III) from the crystalline mineral scorodite ($FeAsO_4.2H_2O$) without reducing As(V), and consequently released Fe(II) and the sorbed As(V) into the solution phase. It is often observed that arsenic mobilisation in flooded soil is accompanied by parallel Fe mobilisation (Masscheleyn et al. 1991; Stroud et al. 2011a; Takahashi et al. 2004; Xu et al. 2008; Yamaguchi et al. 2011). An analogous situation is the release of phosphate by soil flooding even though phosphate does not undergo significant redox transformation in soil. The difference between arsenic and phosphorus is that arsenate released into the solution phase may be reduced to arsenite or even transformed into methylated species (see Sect. 5.2.3), whereas phosphate remains stable. The ratio of As to Fe released varies widely among different soils (Stroud et al. 2011a; Yamaguchi et al. 2011), reflecting the amount of arsenic sorbed by the easily reducible iron oxides/hydroxides. The type and properties (e.g. the degree of crystallinity) of iron oxides/hydroxides have a strong influence on the reductive dissolution and arsenic mobilisation. Ferrihydrite ($Fe(OH)_3$) is amorphous and more easily reduced than other more crystalline iron oxides; however, its dissolution can lead to formation of secondary Fe-bearing minerals (e.g. magnetite, Fe_3O_4), which adsorb the arsenic initially released from the dissolution of ferrihydrite, leading to temporal retention of arsenic until after further development of anaerobic conditions (Tufano and Fendorf 2008; Tufano et al. 2008). In contrast, goethite (α-FeOOH) is more resistant to reductive dissolution, but its dissolution is associated with minimal formation of secondary mineral phases and therefore no temporal retention of arsenic.

Secondly, adsorbed arsenate may be reduced to arsenite in the soil solid phase and, because arsenite is less strongly adsorbed, it has a greater tendency to partition into the solution phase (Takahashi et al. 2004; Tufano et al. 2008; Yamaguchi et al. 2011). Determination of arsenic speciation in the soil solid phase is more difficult than in the solution phase, but has been attempted in several studies using either synchrotron radiation based techniques such as X-ray absorption near edge spectrometry (XANES) (Takahashi et al. 2004; Yamaguchi et al. 2011) or arsenic oxidation state-specific isotope dilution (Hamon et al. 2004). Yamaguchi et al. (2011) showed that As(V) was predominant in the soil solid phase with As(III) accounting for only about 15% in two paddy soils before flooding. After flooding and as Eh decreased, the proportion of As(III) increased to 60–80% at Eh of −100 mV. In comparison, arsenic speciation in the solution phase was dominated by As(III). When the soils were γ-irradiated, reduction of arsenate to arsenite was suppressed even though the irradiation treatment did not totally sterilise the soils. These results illustrate the important role of microbes in mediating arsenate reduction in both the solid and solution phases in soil. Furthermore, As(III) sorbed on the solid phase is much more easily desorbed into the solution phase than As(V). The distribution coefficient (solution phase concentration/solid phase concentration) for As(III) increases rapidly with increasing pH from 5.5 to 7, whereas for As(V) the increase is apparent only at pH >7 (Yamaguchi et al. 2011). In a similar study, Takahashi et al. (2004) estimated that the solid phase arsenate reduction followed by increased arsenite distribution to the solution phase contributed to at least half of the observed arsenic mobilisation upon soil flooding. In a study with ferrihydrite coated sand which was loaded with arsenate, Kocar et al. (2006) showed that arsenate reduction by the bacterium *Bacillus benzoevorans* enhanced the elution of arsenic mainly as arsenite. Based on studies of iron oxide/hydroxide coated sands which were inoculated with bacterium strains capable of Fe(III) and/or As(V) reduction, Tufano et al. (2008) concluded that As(V) reduction followed by enhanced As(III) partition into the solution phase played a dominant role in arsenic mobilisation.

Apart from the two main mechanisms described above, other factors may also contribute to arsenic mobilisation in flooded soils. These include pH increases as a result of soil flooding favouring arsenite and, to a lesser extent, arsenate distribution to the solution phase, and increased concentrations of phosphate and dissolved organic matter (DOM) in the soil solution replacing adsorbed arsenate or arsenite into the solution. In contrast, pH may decrease in alkaline soils after flooding (see Sect. 5.1) and this could result in decreased arsenic distribution to the solution phase in the case where Eh remained above the As(V)-As(III) boundary (Hamon et al. 2004).

5.2.1.3 Arsenate Reduction

From the preceding section, it is clear that arsenate reduction is the key to arsenic mobilisation upon soil flooding. Arsenate reduction may take place either biotically or abiotically; the latter is thermodynamically possible when the reduction is coupled with the oxidation of sulphide, ferrous Fe^{2+}, H_2 gas or reduced organic acid (Inskeep et al. 2002). The abiotic reduction of arsenate by dissolved sulphide is very slow at near neutral pH, but becomes more significant in acidic media with the reduction rate being 300 times faster at pH 4 than at pH 7 (Newman et al. 1997; Rochette et al. 2000). The product arsenite is complexed by sulphide; but precipitation of As_2S_3 (orpiment) may occur when the concentration ratio of S:As is high (Rochette et al. 2000). In paddy soils where pH tends to converge to near neutrality after flooding (see Sect. 5.1) and arsenate reduction generally takes place before sulphate reduction, the abiotic reduction of arsenate by dissolved sulphide is probably unimportant.

Biotic reduction of arsenate mediated by microbes is most likely the dominant process in soil (Jones et al. 2000). Microbial isolates from sediments and soils capable of reducing arsenate have been reported widely (Ahmann et al. 1994; Jones et al. 2000; Macur et al. 2001, 2004; Newman et al. 1997). In fact, bacteria capable of either oxidising As(III) or reducing As(V) coexist and are ubiquitous in soil environments (Macur et al. 2004). There are two main pathways of microbial arsenate reduction: (1) dissimilatory reduction in which arsenate is used as a terminal electron acceptor during anaerobic respiration with a carbon source (e.g. lactate or acetate) as the electron donor (Ahmann et al. 1994; Newman et al. 1997), and (2) detoxification mechanism by which arsenate is reduced by arsenate reductase enzymes such as ArsC and the product arsenite is extruded out of cell via an efflux carrier or an efflux pump (Bhattacharjee and Rosen 2007; Inskeep et al. 2002). Dissimilatory reduction is mediated by a few microbial species and occurs only under anaerobic conditions, whereas the detoxification pathway occurs under both aerobic and anaerobic conditions and is carried out by diverse genera of microorganisms. It is likely that the second pathway plays a more important role in the paddy soil environment.

As mentioned above, the As_2S_3 precipitate may form when the sulphide concentration is sufficiently high. This precipitation may result directly from dissimilatory reduction of both arsenate and sulphate by anaerobic microbes (Newman et al. 1997), from abiotic reduction of arsenate by sulphide (Rochette et al. 2000), or from the reaction between arsenite and sulphide which are produced by separate processes. A consequence of this precipitation is decreased arsenic solubility under highly anoxic conditions, which may present a possible strategy to immobilise arsenic under wetland environments (Inskeep et al. 2002). In some studies, initial arsenite mobilisation after soil flooding was followed by a sharp decrease in the soluble arsenic concentration with unknown reasons (Onken and Hossner 1996); perhaps this was caused by the precipitation of arsenite-sulphide phases. Further studies are needed to evaluate the importance of this process in submerged paddy soils.

5.2.1.4 Arsenite Oxidation

When paddy water is drained away, as is often the practice during the middle-late stage of rice grain filling, the reactions described above are reversed, resulting in the oxidation of Fe(II) and arsenite and a sharp decrease in the soluble arsenic concentration (Arao et al. 2009; Li et al. 2009; Reynolds et al. 1999). Arsenite can be oxidised via abiotic or biotic processes. While arsenite is thermodynamically unstable in the oxygenated environment, its oxidation by O_2 is very slow in aqueous solution with a half life of about 1 year (Inskeep et al. 2002). In contrast, oxidation of arsenite by manganese oxides (e.g. birnessite, δ-MnO_2) is very fast with a half life of less than 1 h (Oscarson et al. 1981, 1983). Other abiotic reactions include oxidation of arsenite by H_2O_2 or free hydroxyl radicals that are generated from the Fenton reaction; these reactions may be particularly relevant in surface waters exposed to UV light (Inskeep et al. 2002).

Microorganisms capable of oxidising arsenite are widespread in soil and sediment (Cai et al. 2009; Macur et al. 2004). Arsenite-oxidising microorganisms are diverse; some are chemolithoautotroph, which gain energy from the oxidation of arsenite (Garcia-Dominguez et al. 2008; Santini et al. 2000), while others may oxidise arsenite as a mechanism of detoxification or for unknown reasons (Anderson et al. 1992). Inskeep et al. (2007) reported more than 160 DNA sequences encoding the AroA-like arsenite oxidase proteins from soil, sediment and geothermal microbial mat. It is clear that microbes play important roles in both the reduction of arsenate and oxidation of arsenite in soil.

5.2.1.5 Impact of Soil Flooding on Arsenic Bioavailability to Rice

Due to the reasons described above, soil redox conditions can be expected to have a profound influence on arsenic bioavailability to rice. The bioavailable pool of arsenic in soil can be measured by the *L*-value using the isotope dilution technique, which was originally developed by Larsen (1969) to measure available phosphorus in soil. To measure arsenic *L*-value, a small quantity of radioactive arsenic (e.g. ^{73}As, in the form of arsenate) is introduced to soil, which redistributes itself among the solution and the exchangeable phases and undergoes redox reactions in the same way as the stable isotope ^{75}As naturally occurring in the soil. Plants (e.g. rice) are grown in the soil to take up both ^{73}As and ^{75}As, and the arsenic *L*-value can be calculated from the radioactivity of ^{73}As added to the soil and that recovered in the plants, and the amount of total arsenic in the plants (Stroud et al. 2011a). In five paddy soils from Bangladesh, India and China, the arsenic *L*-values represent 6–28% of the total arsenic in the soils (Fig. 5.5). Compared with aerobic conditions, flooding doubled the arsenic *L*-value in most soils, clearly indicating increased bioavailable pool of arsenic (Stroud et al. 2011a).

Soil redox conditions also strongly influence arsenic uptake by rice (Arao et al. 2009; Li et al. 2009; Marin et al. 1993; Xu et al. 2008). In a pot experiment, Xu et al. (2008) showed that rice grown under the anaerobic (flooded) conditions contained 7–35 times more arsenic in straw and 10–15 times more arsenic in grain than that under the aerobic conditions. The dramatic differences were observed in both the treatment without arsenic amendment (i.e. only indigenous arsenic in the soil) and those with an addition of arsenate or arsenite. Further studies showed that introducing a period of aerobic soil conditions during the entire rice growth cycle was effective in decreasing arsenic uptake by rice (Arao et al. 2009; Li et al. 2009). These studies highlight the potential of using water management to prevent excessive arsenic accumulation by rice (see Chap. 7).

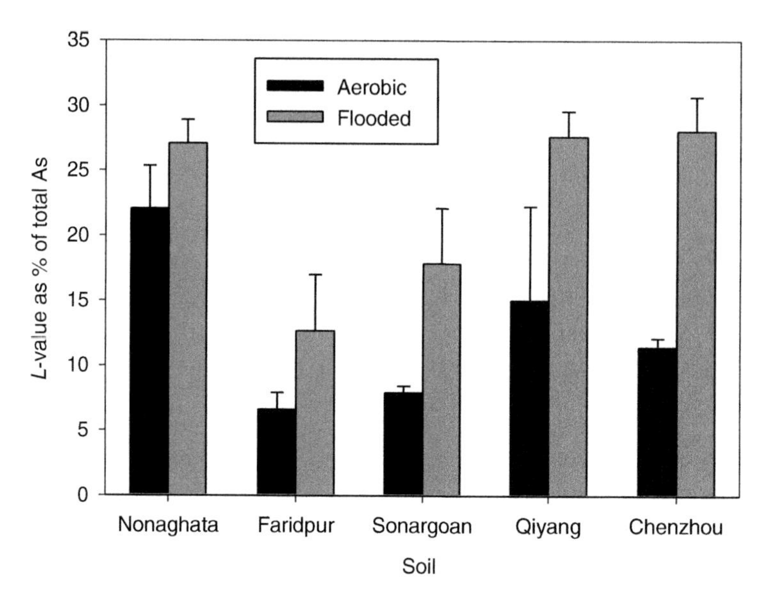

Fig. 5.5 Effect of soil water conditions on arsenic *L*-values (reported as % of total As) in five paddy soils. Nonaghata is from India, Faridpur and Sonargoan are from Bangladesh, and Qiyang and Chenzhou are from China (Redrawn with permission from Stroud et al. (2011a). Copyright (2011) American Chemical Society)

5.2.2 Arsenic Adsorption and Desorption

5.2.2.1 Mechanisms of Adsorption

Adsorption and desorption are key processes controlling the partition of arsenic between the solid and solution phases. In general, arsenate is strongly adsorbed by most mineral constituents of soils, such as various oxides/hydroxides of iron and aluminium, aluminosilicate clay minerals and manganese oxides, whereas arsenite exhibits a limited affinity for most soil minerals except iron oxides/hydroxides (Fendorf et al. 2008). Amorphous iron oxides/hydroxides adsorb much more arsenate or arsenite than their crystalline counterparts because of larger specific surface areas (Dixit and Hering 2003). The content of iron oxides/hydroxides extractable by dithionite-citrate-bicarbonate in soils correlates significantly with the capacity of arsenate adsorption, suggesting an important role of these minerals in arsenic adsorption in soil (Yang et al. 2002).

An arsenic species may interact with a surface functional group(s) on the adsorbent forming a stable molecular entity, which is called a surface complex. There are two types of surface complexes, either outer-sphere or inner-sphere, depending on whether there is a water molecule present between the surface functional group and the bound ion or molecule (yes and no for the two respective complexes) (Sparks 2003).

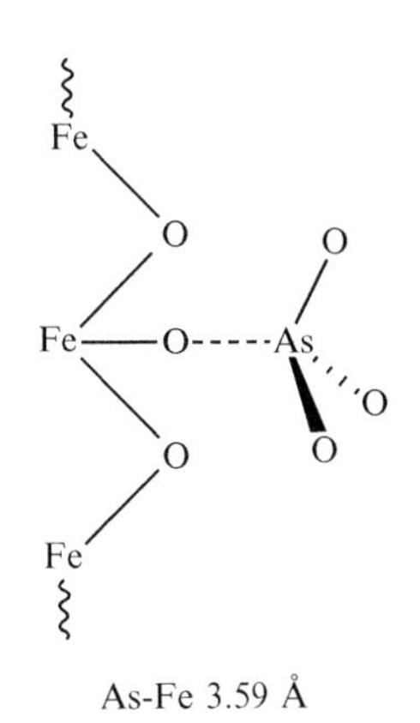

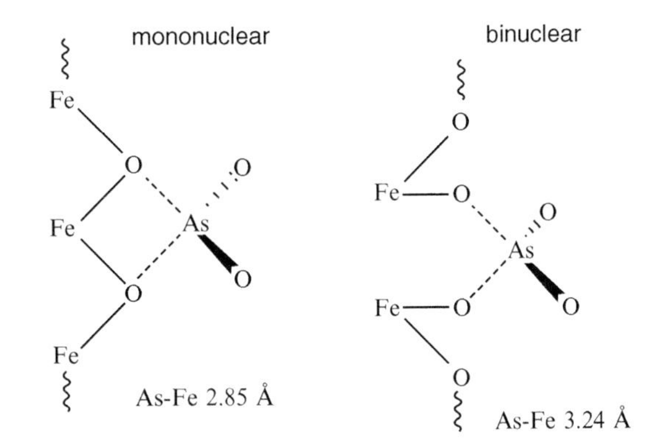

Fig. 5.6 Schematic diagram of the surface structure of arsenate adsorbed on goethite (**a**). Monodentate complex. (**b**) Bidentate complex (Reprinted with permission from Fendorf et al. (1997). Copyright (1997) American Chemical Society)

The strength of adsorption is stronger with the formation of inner-sphere complexes than the outer-sphere complexes. Furthermore, inner sphere complexes can be monodentate (the adsorbate is bonded to only one oxygen) and bidentate (the adsorbate is bonded to two oxygens) (Sparks 2003) (Fig. 5.6). Arsenate forms inner-sphere surface complexes on iron and aluminium oxides/hydroxides primarily in the bidentate binuclear coordination (Arai et al. 2001; Fendorf et al. 1997; Grossl et al. 1997; Lumsdon et al. 1984; Sun and Doner 1996) (Fig. 5.6); this behaviour is similar to the adsorption of phosphate. The strong adsorption explains why arsenate concentrations in soil solution are generally very low; Wenzel et al. (2002) reported 1–4 μg l^{-1} (13–53 nM) in moderately contaminated soils under aerobic conditions and 171 μg l^{-1} (2.3 μM) in a highly contaminated soil (2,250 mg As kg^{-1}), while Hamon et al. (2004) reported 290 μg l^{-1} (3.9 μM) in an acidic soil containing 4,770 mg kg^{-1} total arsenic and 1.4 mg l^{-1} (18.7 μM) in a contaminated alkaline soil (pH 8.3, total As 725 mg kg^{-1}). The high concentration of soluble arsenic in the latter was due to decreased arsenate adsorption at high pH (see below).

Iron oxides/hydroxides provide the most important sorption sites for arsenite (Fendorf et al. 2008). Arsenite adsorbs to iron oxides/hydroxides forming mostly inner-sphere surface complexes in the bidentate binuclear coordination, with some bidentate mononuclear and monodentate coordination also present depending on the type of iron oxides/hydroxides (Manning et al. 1998; Ona-Nguema et al. 2005; Sun and Doner 1996). Outer-sphere adsorption of arsenite arising from electrostatic interactions and hydrogen bonding may also occur (Fendorf et al. 2008; Goldberg and Johnston 2001).

Methylated arsenic species such MMA and DMA are also adsorbed extensively by iron oxides/hydroxides (Lafferty and Loeppert 2005). Extended X-ray absorption fine structure (EXAFS) studies show that arsenic atoms in MMA and DMA likely form bidentate binuclear inner-sphere complexes on the surface of goethite (Shimizu et al. 2011a). Formation of outer-sphere complexes via electrostatic interactions or hydrogen bonding is also possible, particularly for DMA (Shimizu et al. 2011a). The adsorption of both species in soil correlates with the content of Fe/Al oxides/hydroxides, and there is a strong co-localisation of arsenic and iron in micro-scale mapping of soil particles using synchrotron based X-ray fluorescence, indicating that Fe/Al oxides/hydroxides are the main sorbents for the methylated arsenic species in soil (Shimizu et al. 2011a, b). Compared with inorganic arsenic species, increased methyl substitution of the hydroxyl group results in both decreased adsorption and increased ease of release from the iron oxide surface (Lafferty and Loeppert 2005). The adsorption of MMA in soil is greater than that of DMA, whereas the opposite is true for desorption (Shimizu et al. 2011a). The alumina (α-Al_2O_3) mineral has a much smaller capacity for the adsorption of MMA and DMA than for arsenate (Xu et al. 1991).

The adsorption of arsenite and arsenate by iron oxides/hydroxides exhibits different responses to changing pH. Increasing pH markedly decreases arsenate adsorption, whereas arsenite adsorption is relatively insensitive to pH changes with the adsorption plateau occurring in the neutral – alkaline range (Dixit and Hering 2003; Raven et al. 1998) (Fig. 5.7). Thus, adsorption of arsenite can be greater than that of arsenate except under the acidic conditions. However, the adsorbed arsenite is more labile than arsenate and therefore more prone to desorption. This is nicely illustrated by column leaching studies with iron oxide/hydroxide coated sands that were preloaded with either arsenate or arsenite. Despite being adsorbed to a greater extent than arsenate, arsenite desorbed more rapidly and extensively from all iron oxides, suggesting a weaker binding of arsenite than arsenate (Kocar et al. 2006; Tufano and Fendorf 2008). This is also true for paddy soils, with arsenite exhibiting a much larger solution/solid phase ratio than arsenate (Yamaguchi et al. 2011).

5.2.2.2 Effects of Competitive Ligands

Adsorption of arsenate or arsenite is subject to competition by other ligands such as phosphate, silicate, carbonate and organic acids (Zhang and Selim 2008). In particular, the presence of phosphate strongly suppresses the adsorption of both arsenic

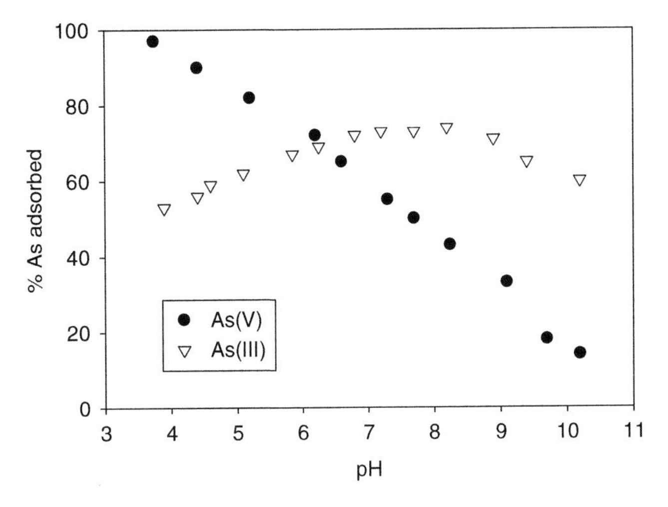

Fig. 5.7 Comparison of arsenate and arsenite sorption edges on hydrous ferric oxide. Initial arsenic concentration was 50 μM (Redrawn with permission from Dixit and Hering (2003). Copyright (2003) American Chemical Society)

species on iron oxides/hydroxides (Dixit and Hering 2003; Jain and Loeppert 2000; Manning and Goldberg 1996). The effect of phosphate on arsenate adsorption was greater at high pH than at low pH, whereas the opposite trend was observed for arsenite (Jain and Loeppert 2000). Phosphate inhibited arsenate adsorption more on gibbsite (γ-$Al(OH)_3$) than on goethite (α-FeOOH), whereas the reverse was observed with regard to the effect of arsenate on phosphate adsorption (Violante and Pigna 2002). Additions of phosphate fertilisers can lead to increased arsenate desorption and hence its mobility in soil (Peryea 1991). In a pot experiment with flooded soils, additions of phosphate fertilisers were found to increase arsenic accumulation by rice, probably as a result of increased desorption from the soil solid phase (Hossain et al. 2009). Although phosphate also inhibits arsenate uptake by plant roots (see Chap. 6), in flooded soil the arsenate displaced by phosphate from the solid phase is likely to be reduced to arsenite, which is not taken up by roots via the phosphate transport pathway. Additions of phosphate enhanced arsenate reduction to arsenite under anaerobic conditions, possibly because of increased arsenate desorption which subsequently facilitated the reduction process in the solution phase (Reynolds et al. 1999). For paddy rice roots coated with iron plaque (see Chap. 6), increasing phosphate availability can also enhance desorption of arsenate and arsenite adsorbed on the plaque, possibly leading to increased arsenic availability for plant uptake.

Silicic acid is another natural ligand in soils that can compete with the adsorption of arsenate and arsenite, especially the latter. Waltham and Eick (2002) showed that the presence of 1 mM silicic acid decreased arsenite adsorption on goethite by 40%. Li et al. (2009) found that an addition of sparingly soluble SiO_2 gel to a flooded soil considerably increased the concentrations of arsenite and, to a lesser extent, arsenate in the soil solution. Unlike phosphate additions, increased Si availability decreased arsenite accumulation by rice by inhibiting arsenite uptake via the silicic acid/arsenite transporters (see Chap. 6).

A consequence of soil flooding is the accumulation of carbonate and bicarbonate in the soil solution (Kirk 2004). Carbonate was found to either increase or decrease arsenate adsorption on hematite (Fe_2O_3) depending on the available surface sites, initial arsenate concentrations and reaction times, with generally small effects (Arai et al. 2004). However, the study of Arai et al. (2004) was conducted with a high arsenate concentration (~0.4 mM) and under normal (i.e. low) CO_2 pressure ($P_{CO2} = 10^{-3.5}$ atm). Stachowicz et al. (2007) used a concentration of bicarbonate (10 mM) that is more representative of those found in the groundwater of Bangladesh. They found that carbonate/bicarbonate had a small effect in suppressing arsenate adsorption on goethite, and a larger effect on arsenite adsorption with an up to tenfold increase in the solution concentration of arsenite in the presence of carbonate/bicarbonate at near neutral pH. The results from Radu et al. (2005) are similar, showing small increases in arsenate mobility in a column of iron oxide coated sand in response to increasing CO_2 partial pressure and a greater effect on arsenite mobility. Overall, the effect of increased carbonate/bicarbonate concentrations on arsenic mobility is noticeable, especially for arsenite, but relatively small compared to those of phosphate and silicic acid.

Dissolved organic matter (DOM) in soil may compete with arsenic for adsorption sites and thus increase arsenic mobility. Xu et al. (1988, 1991) showed that fulvic acid decreased the adsorption of arsenate and arsenite on alumina. Similarly, peat humic acid and a Suwannee River fulvic acid decreased the adsorption of arsenate and arsenite on goethite (Grafe et al. 2001). Citric acid inhibited arsenate adsorption on ferrihydrite but not on goethite, and decreased arsenite adsorption on both minerals (Grafe et al. 2001, 2002). Because of the competition for adsorption, DOM can significantly mobilise arsenate and arsenite adsorbed on iron oxides and soils (Bauer and Blodau 2006), resulting in increased mobility and transport in column leaching studies (Sharma et al. 2011). The mobilisation effect of DOM is smaller than that of phosphate, suggesting that DOM mainly exchanges with weakly sorbed arsenic (Bauer and Blodau 2006). Dissolved organic matter generally increases with soil flooding (Li et al. 2010; Xu et al. 2008), which may contribute to increased arsenic mobility.

DOM may also influence arsenic mobility and bioavailability through complexation. It has been shown that DOM can complex arsenate and arsenite either directly or indirectly through a metal (e.g. Fe) bridge (Buschmann et al. 2006; Liu et al. 2011; Ritter et al. 2006; Sharma et al. 2010) (Fig. 5.8). Buschmann et al. (2006) found that arsenate was more strongly bound by DOM than arsenite. Compared with DOM, Fe(III)-DOM exhibited a greater ability to complex arsenic (Bauer and Blodau 2009; Liu et al. 2011; Sharma et al. 2010). Depending on the experimental conditions (e.g. As:DOM ratio, pH), the conditional stability constants of the As(III)/As(V)-DOM or As(III)/As(V)-Fe-DOM complexes are in the range of $10^{0.5-6.5}$ (Liu and Cai 2010; Liu et al. 2011; Sharma et al. 2010; Warwick et al. 2005). The percentage of soluble arsenic complexed by DOM or Fe-DOM varies widely, from <1% to 70%, among different studies, reflecting a strong

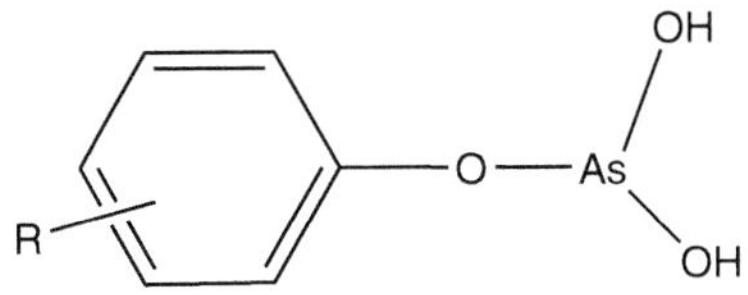

DOM-As(III)

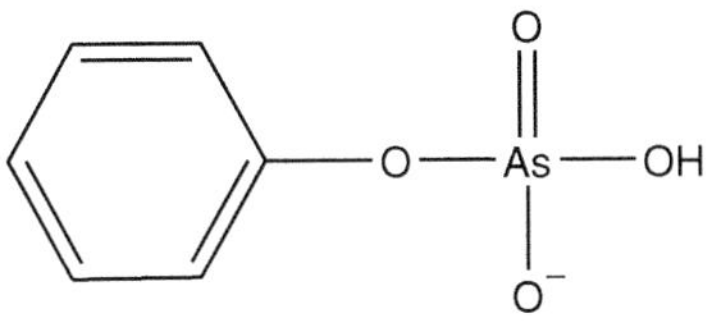

DOM-As(V)

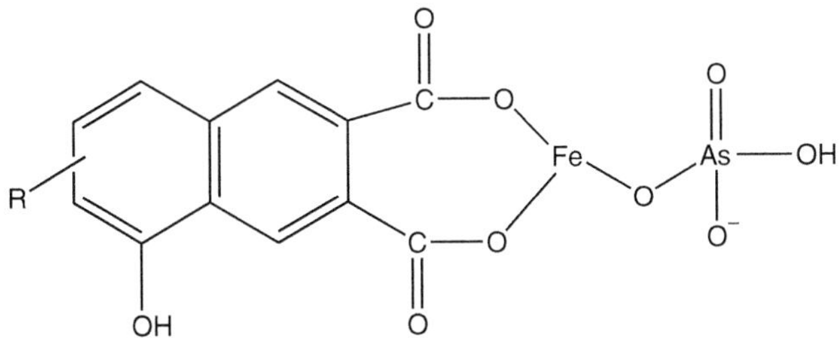

DOM-Fe-As(V)

Fig. 5.8 Possible DOM-As complexes (Redrawn with permission from Buschmann et al. (2006) and Liu et al. (2011). Copyright American Chemical Society)

influence of the experimental conditions (Buschmann et al. 2006; Liu and Cai 2010; Ritter et al. 2006; Sharma et al. 2010). It should be noted that many of the studies referred to here used commercial humic acids. The extent and strength of As-DOM complexation in paddy soils and how the complexation influences arsenic mobility and bioavailability to rice plants remain unclear.

5.2.3 Arsenic Biomethylation, Volatilisation and Demethylation

5.2.3.1 Organic Arsenic Species in Soils

Whilst arsenic in soil is predominantly inorganic, organic arsenic species have been detected in some soils. Takamatsu et al. (1982) used 1 M HCl to extract a range of contaminated paddy or orchard soils and reported 2–69 μg As kg^{-1} of DMA and

0–88 μg As kg^{-1} of MMA, as well as trace levels of two unidentified organic arsenic species. Moreover, they found that paddy soils contained more organic arsenic than upland orchard soils. Compared with inorganic arsenic, the concentrations of DMA and MMA were 3–4 orders of magnitude smaller. Takamatsu et al. (1982) also monitored changes in different arsenic species in a paddy soil during a rice growing season. As expected, the relative proportion of arsenate decreased and that of arsenite increased after flooding. The pattern of the DMA dynamics was rather similar to that of arsenite, suggesting that flooding was conducive to arsenic methylation, whilst MMA remained relatively stable. In a pot study of flooded rice irrigated with arsenate-containing water, arsenite, arsenate, DMA and, occasionally, MMA, were detected in the soil solution (Abedin et al. 2002).

Organic arsenic species are found in other soils. Geiszinger et al. (2002) reported small concentrations of trimethylarsine oxide and arsenobetaine in a heavily contaminated soil due to arsenopyrite (FeAsS) mineralisation. Huang and Matzner (2006) identified the presence of MMA, DMA, trimethylarsine oxide, arsenobetaine and two unknown organic arsenic species, at concentrations up to 14 μg As kg^{-1}, in a peaty forest soil. In the same study, MMA, DMA, arsenobetaine and trace levels of trimethylarsine oxide and tetramethylarsonium ion were detected in the pore water samples collected at a time when microbial activity was high and the redox conditions were reducing. However, DMA disappeared from the pore water when the peaty soil became aerobic (Blodau et al. 2008).

Preservation of soil solutions from anaerobic paddy soils is critical for arsenic analysis and is often done with an addition of a chelating agent (e.g. EDTA) or an acid. Without preservation, soluble ferrous iron is rapidly oxidised forming precipitates of ferric hydroxides, which then adsorb arsenic from the solution phase, resulting in erroneously low concentrations of soluble As. Whilst a sufficiently high concentration of EDTA appeared to preserve inorganic arsenic species well, methylated arsenic species were rarely detectable in the EDTA-treated soil solutions (Stroud et al. 2011b; Xu et al. 2008). In contrast, the use of HCl was found to maintain DMA and MMA in anaerobic soil solutions better (Liu et al., unpublished).

It is clear that organic arsenic species can be present in soils, although usually at very small concentrations. Furthermore, anaerobic conditions and high organic matter content may favour arsenic methylation.

5.2.3.2 The Pathway of Arsenic Biomethylation

In the late nineteenth century, the Italian physician Bartolomeo Gosio first demonstrated that some fungi were able to convert inorganic arsenic into a volatile and toxic arsenic gas with a garlic smell. The volatile compound, named “Gosio” gas, was identified as trimethylarsine [CH_3As] by Challenger and his co-workers in the 1930s–1940s (Bentley and Chasteen 2002; Cullen and Reimer 1989). Challenger et al. proposed a mechanism for the formation of trimethylarsine via repeated steps of reduction of As(V) to As(III) followed by transfer of the methyl group from a

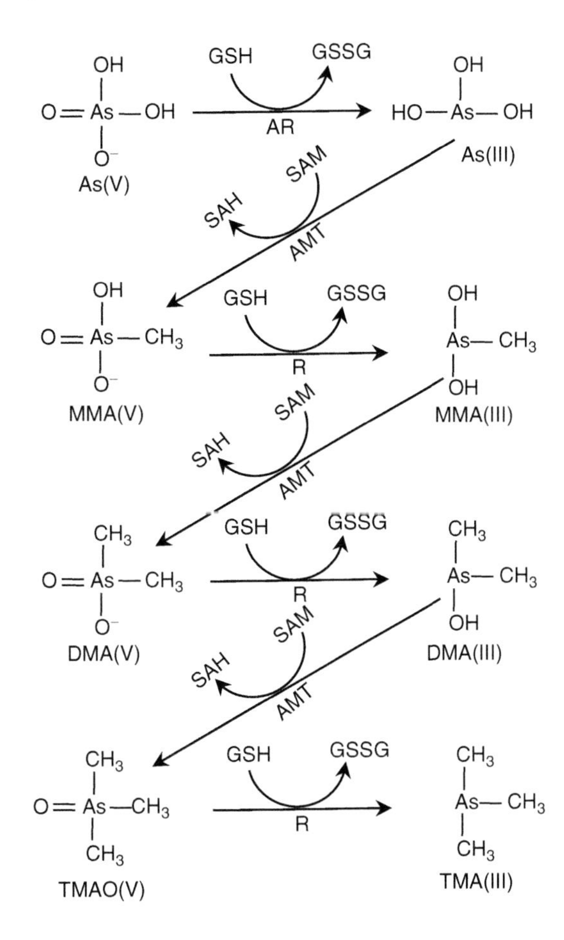

Fig. 5.9 The Challenger pathway of arsenic biomethylation. *GSH* reduced glutathione, *GSSG* oxidised glutathione, *AR* arsenate reductase, *SAM* *S*-adenosylmethionine, *SAH* *S*-adenosylhomocysteine, *R* reductase (Redrawn with permission from Zhao et al. (2010))

donor molecule to the trivalent arsenic (Fig. 5.9) (Challenger 1945). In the Challenger pathway, the methyl transfer step is an oxidative process, hence there is a need for As(V) reduction before the addition of the next methyl group. The methyl donor was later identified to be *S*-adenosylmethionine (SAM), although there is some controversy with regard to whether some bacteria and anaerobic microorganisms use methylcobalamin (vitamin B_{12}) as the methyl donor (Bentley and Chasteen 2002; Cullen and Reimer 1989).

More recently, Hayakawa et al. (2005) proposed an alternative pathway of arsenic methylation in humans. In their proposed model, As(III) and MMA(III) conjugated by glutathione [As(III)-GS_3, MMA(III)-GS_2] are the substrates for methyltransferase, and the products of the reactions are MMA(III)-GS_2 and DMA(III)-GS, respectively. Both MMA(III)-GS_2 and DMA(III)-GS are unstable in solution with a low concentration of reduced glutathione (GSH), and are hydrolysed and oxidised to MMA(V) and DMA(V), respectively. In this scheme, both

MMA(V) and DMA(V) are the end products, rather than the intermediates as in the Challenger pathway.

The ability to biomethylate arsenic is widespread in different organisms, including many fungi, bacteria, archaea, algae, and animals. A wide range of aerobic and anaerobic microorganisms in soil are able to biomethylate arsenic; the latter include methanogenic archaea which are prevalent in anaerobic paddy soil. It has been assumed that higher plants are also able to biomethylate arsenic; however, recent studies have casted serious doubt on this assumption (see Chap. 6). It is probable that plants acquire methylated arsenic from the soil.

In humans and rats, the transfer of the methyl group to As(III) is catalysed by the enzyme As(III) *S*-adenosylmethyltransferase, named Cyt19 or As3MT (Lin et al. 2002; Thomas et al. 2004). Homologues of Cyt19 have been identified in a large number of bacteria, archaea and algae; many of them live in soil (Qin et al. 2006). A subset of these homologous genes are downstream of an *arsR* gene, which encodes the archetypal As-responsive transcriptional repressor that controls expression of *ars* operons; these genes are termed *arsM* and their protein arsM (As(III) S-adenosylmethyltransferase). When *arsM* from the soil bacterium *Rhodopseudomonas palustris* was expressed in an As-sensitive strain of *Escherichia coli*, inorganic arsenic was transformed to a number of methylated intermediates, such as DMA(V) and trimethylarsine oxide, and the end product trimethylarsine which was volatilised into the headspace of the incubation vessels (Qin et al. 2006; Yuan et al. 2008). Biomethylation and subsequent volatilisation of trimethylarsine lessen the cellular burden of arsenic, thus conferring arsenic resistance to the bacterium. For this reason biomethylation is considered to represent an arsenic detoxification mechanism in bacteria and fungi (Qin et al. 2006). This also explains why *arsM* is under the transcriptional regulation of *arsR*. When *R. palustris arsM* was expressed in rice, there was also increased arsenic volatilisation from the transgenic plants, although the extent of arsenic methylation was limited (Meng et al. 2011) (see Chap. 6).

Arsenic methylation appears to be enhanced by anaerobic conditions in soil (Shimizu et al. 2011b). When MMA(V) was added to a soil under anaerobic conditions, MMA(III), DMA(III) and DMA(V) were produced (Shimizu et al. 2011b). In contrast, aerobic soil conditions favour the degradation (demethylation) of MMA(V) and DMA(V) (see Sect. 5.2.3.4).

5.2.3.3 Arsenic Volatilisation

Mestrot et al. (2011; 2009) used a chemo-trap method to collect and quantify gaseous arsines volatilised from soils in microcosm experiments and from paddy fields. Trimethylarsine was found to be the main species, with arsine (AsH_3), monomethylarsine, dimethylarsine also being detected in some cases. Flooding soil and additions of organic manure increased arsenic volatilisation; very little arsenic was emitted from a paddy soil incubated under aerobic conditions (Mestrot et al. 2009). In soil microcosm experiments with a range of soils varying in the level of arsenic

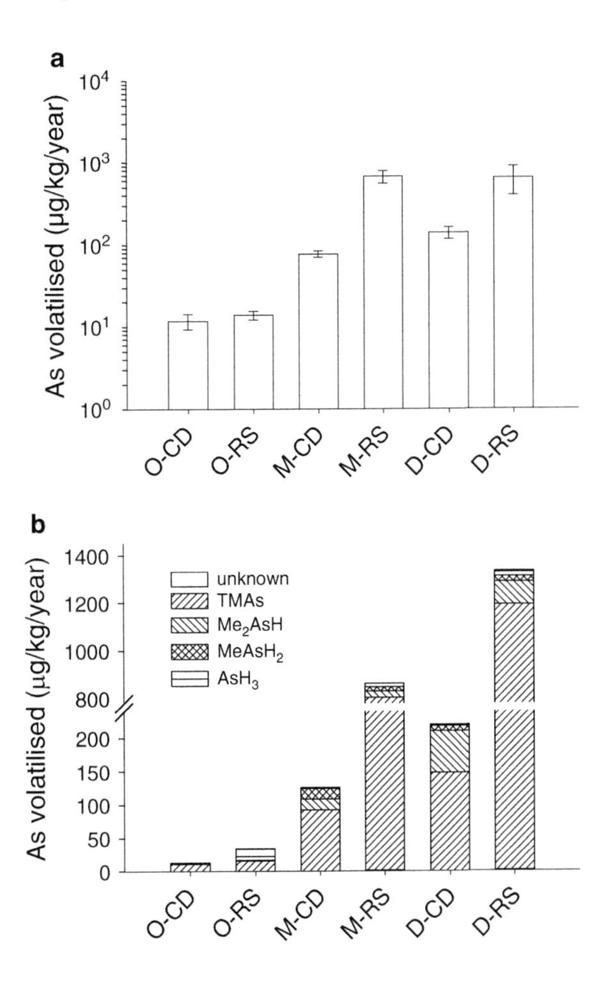

Fig. 5.10 Volatilisation of arsine gas (**a**) and the arsenic species (**b**) from a Bangladeshi paddy soil in a microcosm experiment. *O* no addition of MMA or DMA, *M* addition of MMA, *D* addition of DMA, *CD* +cow dung, *RS* +rice straw (Redrawn with permission from Mestrot et al. (2011). Copyright (2011) American Chemical Society)

contamination, the amount of arsenic volatilised ranged from 0.5 to 70 μg kg^{-1} soil $year^{-1}$, representing 0.00–0.17% of the total soil arsenic (Mestrot et al. 2011). Volatilisation increased when MMA or DMA was added to a Bangladeshi paddy soil, especially when rice straw was also added (Fig. 5.10), suggesting that volatilisation was limited by the activity of arsenic biomethylation in the soil. Soluble arsenic in the soil pore water and dissolved organic C were two significant factors correlating with the flux of arsenic volatilisation. It appears that arsenic biomethylation in soil, and subsequently volatilisation of arsines, increases with the availability of inorganic arsenic in the soil pore water. Flooding and organic manuring both promote the reductive mobilisation of arsenic into the soil pore water (see 5.2.1), thus favouring arsenic biomethylation and volatilisation. Another important factor is the abundance and/or the activity of microorganisms that biomethylate arsenic in soil, but how much their abundance varies among soils and how soil conditions affect their activities remain unclear.

In field measurements, 44 and 240 mg As ha^{-1} $year^{-1}$ were volatilised from a Spanish and a Bangladeshi paddy field, respectively, representing only 0.0003–0.0016% of the total arsenic in the top 10 cm soil (Mestrot et al. 2011). Field measurements of arsenic volatilisation were about 1–2 orders of magnitude lower than those made with laboratory microcosm experiments using small amounts of soils. In general, it can be concluded that volatilisation of arsine gases occurs from paddy fields, but the flux is very small compared with the amount of arsenic in the soil.

5.2.3.4 Demethylation

Methylated arsenic species, either of microbial origin or added to soil as pesticides or herbicides, can be transformed by microorganisms via two pathways: (1) reductive conversion to volatile organo-arsine species (e.g. dimethyl- or trimethylarsine) and emitted from the soil system; and (2) demethylation to produce the end products CO_2 and arsenate. The first pathway predominates under anaerobic conditions, whereas both pathways occur in aerobic soil (Woolson and Kearney 1973). Gao and Burau (1997) showed that demethylation was quantitatively far more important than the evolution of gaseous arsines when DMA was added to an aerobic soil. In a field study, the half-life of the applied DMA and MMA was found to be about 20 days, although these compounds were still detectable in soil 1.5 years after applications (Woolson et al. 1982). Demethylation of DMA was relatively slow in the soil extracts from a forest floor and from an oxic fen soil, with half-lives of 187 and 46 days, respectively (Huang et al. 2007). In comparison, arsenobetaine was rapidly degraded to DMA with half-lives of 4–12 days (Huang et al. 2007).

Some soil bacteria able to demethylate MMA to inorganic arsenic have been isolated, for example two strains of *Pseudomonas putida* from soils heavily contaminated by organic arsenic used in chemical weapon reagents (Maki et al. 2006). In some cases, demethylation is carried out in two sequential steps, firstly reduction of pentavalent MMA(V) to MMA(III) and secondly demethylation of the latter to As(III). These two steps were found to be performed by two different bacteria belonging to *Burkholderia* and *Streptomyces* species, respectively, in golf-course soils in Florida that have received MMA as an herbicide (Yoshinaga et al. 2011).

5.2.4 *Predicting Available Arsenic in Paddy Soil*

It is a well established paradigm that the total concentration of a contaminant in soil does not reflect its bioavailability to organisms (e.g. plants). This is also the case for arsenic in paddy soil (Khan et al. 2010); an exception may be found when a study is conducted within a small area having the same soil type and similar basic soil properties. For example, Hossain et al. (2008) investigated the relationships between the arsenic concentration in rice grain and various measurements of soil arsenic in samples collected from paddy fields within a single shallow tube-well command area (8 ha). They found a significant correlation between grain arsenic concentration and soil total arsenic ($r = 0.815$, $p < 0.001$, $n = 56$), which is stronger than the correlations with either phosphate-extractable arsenic ($r = 0.677$, $p < 0.001$) or oxalate-extractable arsenic ($r = 0.804$, $p < 0.001$). However, when different soil types or different sources of arsenic contamination are encompassed, the relationship between plant arsenic concentration and soil total arsenic is generally poor (Khan et al. 2010; Lu et al. 2009). Phosphate-extractable As, which measures specifically sorbed arsenic in soil, was found to correlate significantly with arsenic concentration in rice grain of both Boro and Aman rice across nine field stations of the Bangladesh Rice Research Institute (Ahmed et al. 2011), and with straw arsenic concentration in a glasshouse study using paddy soils contaminated by different sources of arsenic (Khan et al. 2010). Phosphate-extractable arsenic also predicted the concentration of arsenic mobilised in the soil pore water under flooded conditions well (Stroud et al. 2011a). Multiple regression can provide clues to the main soil factors that influence arsenic uptake. Bogdan and Schenk (2009) found that the concentrations of arsenic in rice grain and straw could be explained by soil total arsenic and the contents of poorly crystalline iron oxides and plant available phosphorus, with the percentage of the variance accounted for by the regression model ranging from 37% to 77%. In the regression models, the effect of iron oxides was negative, consistent with their role in adsorbing arsenic species, whereas the effects of soil total arsenic and available P were positive. The positive effect of P may be due to phosphate replacing adsorbed arsenic on iron oxides/hydroxides in the soil or on the root surfaces (iron plaque), thus increasing the concentration of arsenic in the soil solution. In another study, Williams et al. (2011) showed that grain arsenic concentration in Aman rice could be explained well by the soil arsenic pool measured by DGT (diffusive gradient in thin film) and dissolved organic matter in the pore water ($R^2 = 0.72$, $p < 0.001$, $n = 32$). The coefficient for DOM was negative in the multiple regression model obtained, suggesting that increasing DOM may decrease arsenic uptake by rice.

The difficulties in finding a simple and reliable method to predict arsenic uptake by rice are understandable considering that multiple factors and interactions between them influence arsenic bioavailability and uptake by plants. A key factor that has been discussed extensively in this Chapter is the strong impact of soil redox potential on arsenic mobilisation in paddy soil; yet this strong effect is difficult to quantify using any measurements at a single time point because soil redox potential may fluctuate greatly during the rice growing season. Another factor worthy of a note is that arsenic toxicity in rice may actually decrease arsenic concentration in the grain, further complicating the relationship between soil arsenic measurements and grain arsenic concentration (Khan et al. 2010; Panaullah et al. 2009).

5.3 Impact of Groundwater Irrigation on Arsenic Dynamics in Paddy Fields

In south Asia, two crops of rice are typically grown in a year: the dry season (Boro) rice from December to April and the wet season (Aman) rice from June to November. Boro rice is more productive than Aman rice, producing on average 1.7 times larger grain yield than the latter, and for this reason, its area has quadrupled in Bangladesh since the early 1970s while the area for Aman rice has remained static (http://www.moa.gov.bd/statistics/statistics.htm). Cultivation of Boro rice relies heavily on irrigation; 90% of the Boro rice area in Bangladesh is irrigated, accounting for 72% of the total irrigated crop area in the 2004–2005 season. Groundwater extracted via tube-wells provides the main source of irrigation water covering 74% of the irrigated crop area. Where the groundwater is contaminated with arsenic, as in many cases in the Bengal delta, irrigation can add considerable amounts of arsenic to the paddy environment. In Bangladesh, arsenic input from irrigation amounts to 1,360 tonnes per year (Ali and Jain 2004). Boro rice typically requires irrigation of about 1,000 mm water during the entire growing season (Saha and Ali 2007; Stroud et al. 2011b). At a concentration of 0.2 mg As l^{-1}, the irrigation water adds 2 kg As ha^{-1}, which could increase the total arsenic concentration in the topsoil by 1.5 mg kg^{-1} if all of the arsenic added is retained in the top 0–15 cm soil having a bulk density of 1.3 g cm^{-3}. However, as discussed below, the dynamics of arsenic is spatially and temporally heterogeneous, and re-mobilisation of arsenic from paddy soil to the flood water during the monsoon season results in a significant attenuation of arsenic in the paddy environment.

Groundwater extracted from the shallow tube-wells in the Bengal Delta generally contains arsenite as the dominant arsenic species, reflecting the reducing environment of the aquifers (Roberts et al. 2007; Stroud et al. 2011b). After irrigation, ferrous iron and arsenite are oxidised rapidly leading to formation and settling of As-bearing hydrous ferric oxide aggregates and sharp decreases in their concentrations in the standing water across the paddy field from the irrigation inlet (Roberts et al. 2007). Both arsenate and arsenite in the paddy standing water can also be adsorbed by soil minerals. In one study, total arsenic concentration in the field standing water was found to be approximately threefold higher near the irrigation inlet than at the opposite corner of the field 70 m away from the inlet shortly after an irrigation event (Roberts et al. 2007).

Consistent with the arsenic dynamics in the field standing water following irrigation is the distinct spatial pattern of declining total arsenic concentration in the surface soil with the distance from the irrigation inlet (Dittmar et al. 2007; Hossain et al. 2008; Lu et al. 2009; Panaullah et al. 2009; Saha and Ali 2007; Stroud et al. 2011a). Such spatial pattern is most pronounced in the paddy fields that have been irrigated with high arsenic groundwater; where tube-well water contains low levels of arsenic, no clear gradient in soil arsenic concentration is apparent. An example is given in Fig. 5.11 for a paddy field in Faridpur, Bangladesh, which had been irrigated with shallow tube-well water containing about 0.2 mg As L^{-1} for 18 years

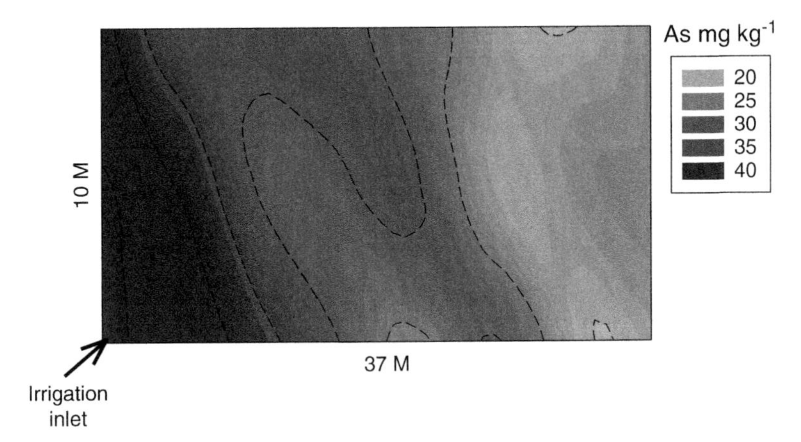

Fig. 5.11 Spatial variation in topsoil (0–15 cm) arsenic concentration in a paddy field irrigated with shallow tube-well water containing ~0.2 mg As L^{-1}, in the Faridpur district, Bangladesh (Unpublished data based on the study of Stroud et al. (2011b))

before soil sampling. In more extreme cases, within-field variation in topsoil arsenic concentration was up to sevenfold (Panaullah et al. 2009). Arsenic distribution not only varies laterally across the field, but also vertically down the soil profile. Dittmar et al. (2007) showed a strong arsenic enrichment in the top 1 cm of the soil (149 mg As kg^{-1}), followed by a second peak at the 5–10 cm depth (54 mg As kg^{-1}) that may be caused by a high density of rice roots accumulating arsenic on the iron plaque. Arsenic concentrations were much smaller and relatively stable between the 30–70 cm depth (4–11 mg As kg^{-1}).

Whilst some accumulation of arsenic in the topsoil is evident, albeit heterogeneously across the field, mass-balance calculations indicate substantial losses of arsenic from the paddy systems (Dittmar et al. 2007; 2010). Analyses of soil arsenic concentration in successive years often show no or only small year-on-year increases after irrigation of As-contaminated water (Dittmar et al. 2007; Lu et al. 2009; Saha and Ali 2007).There are several potential pathways of arsenic losses, including plant uptake, leaching of arsenic down the soil profile, mobilisation of arsenic into the paddy flood water which drains away horizontally, and volatilisation. Plant uptake of arsenic is very small (typically <50 g ha^{-1}) compared with the amount of arsenic added in the irrigation water (2 kg ha^{-1} from 1,000 mm water containing 0.2 mg As L^{-1}). Arsenic volatilisation is also negligible compared with the input (see Sect. 5.2.3.3). Leaching of some arsenic down the soil profile is possible, but not substantial according to studies on the vertical distribution of arsenic concentration (Dittmar et al. 2007; 2010; Norra et al. 2005; Saha and Ali 2007), probably because the presence of a plough pan below the plough layer effectively prevents vertical drainage. In a study using intact soil columns (40 cm depth) irrigated with As-contaminated water, Khan et al. (2009) showed that arsenic uptake by four successive rice crops in 2 years and cumulative arsenic leaching accounted for only 1–3%

and 1–5%, respectively, of the arsenic added in the irrigation water, with the majority (80–90%) of arsenic being retained in the soils.

The most important pathway of arsenic losses is through lateral drainage of flood water from the paddy field during the monsoon season when a substantial amount of arsenic is remobilised from the soil due to the development of anaerobic conditions and this diffuses into the flood water. When the flood water recedes, the mobilised arsenic is transferred from the paddy field to the surrounding rivers. Roberts et al. (2010) estimated that this pathway removed 13–62% of the arsenic added to soils through irrigation each year. With 21% of the land area in Bangladesh being affected annually by monsoon floodwater exceeding 0.9 m depth, arsenic removal through laterally receding floodwater is a common pathway of arsenic attenuation in the paddy system in the Bengal delta region (Roberts et al. 2010). Indeed, it has been observed that the arsenic concentration in soil builds up after irrigation in the Boro rice season, but decreases following the monsoon season (Dittmar et al. 2007; 2010; Saha and Ali 2007). Where paddy fields are not frequently inundated by monsoon flooding, irrigation of As-contaminated groundwater is likely to result in more pronounced accumulation of arsenic in the soils.

References

Abedin MJ, Feldmann J, Meharg AA (2002) Uptake kinetics of arsenic species in rice plants. Plant Physiol 128:1120–1128

Ahmann D, Roberts AL, Krumholz LR, Morel FMM (1994) Microbe grows by reducing arsenic. Nature 371:750

Ahmed ZU, Panaullah GM, Gauch H, McCouch SR, Tyagi W, Kabir MS, Duxbury JM (2011) Genotype and environment effects on rice (*Oryza sativa* L.) grain arsenic concentration in Bangladesh. Plant Soil 338:367–382

Ali I, Jain CK (2004) Advances in arsenic speciation techniques. Int J Environ Anal Chem 84:947–964

Anderson GL, Williams J, Hille R (1992) The purification and characterization of arsenite oxidase from *Alcaligenes faecalis*, a molybdenum-containing hydroxylase. J Biol Chem 267:23674–23682

Arai Y, Elzinga EJ, Sparks DL (2001) X-ray absorption spectroscopic investigation of arsenite and arsenate adsorption at the aluminum oxide-water interface. J Colloid Interface Sci 235:80–88

Arai Y, Sparks DL, Davis JA (2004) Effects of dissolved carbonate on arsenate adsorption and surface speciation at the hematite-water interface. Environ Sci Technol 38:817–824

Arao T, Kawasaki A, Baba K, Mori S, Matsumoto S (2009) Effects of water management on cadmium and arsenic accumulation and dimethylarsinic acid concentrations in Japanese rice. Environ Sci Technol 43:9361–9367

Bartlett RJ, James BR (1993) Redox chemistry of soils. Adv Agron 50:151–208

Bauer M, Blodau C (2006) Mobilization of arsenic by dissolved organic matter from iron oxides, soils and sediments. Sci Total Environ 354:179–190

Bauer M, Blodau C (2009) Arsenic distribution in the dissolved, colloidal and particulate size fraction of experimental solutions rich in dissolved organic matter and ferric iron. Geochim Cosmochim Acta 73:529–542

Bentley R, Chasteen TG (2002) Microbial methylation of metalloids: arsenic, antimony, and bismuth. Microbiol Mol Biol Rev 66:250–271

Bhattacharjee H, Rosen BP (2007) Arsenic metabolism in prokaryotic and eukaryotic microbes. In: Nies DH, Silver S (eds) Molecular microbiology of heavy metals. Springer, Berlin, pp 371–406
Blodau C, Fulda B, Bauer M, Knorr KH (2008) Arsenic speciation and turnover in intact organic soil mesocosms during experimental drought and rewetting. Geochim Cosmochim Acta 72:3991–4007
Bogdan K, Schenk MK (2009) Evaluation of soil characteristics potentially affecting arsenic concentration in paddy rice (*Oryza sativa* L.). Environ Pollut 157:2617–2621
Borch T, Kretzschmar R, Kappler A, Van Cappellen P, Ginder-Vogel M, Voegelin A, Campbell K (2010) Biogeochemical redox processes and their impact on contaminant dynamics. Environ Sci Technol 44:15–23
Buschmann J, Kappeler A, Lindauer U, Kistler D, Berg M, Sigg L (2006) Arsenite and arsenate binding to dissolved humic acids: influence of pH, type of humic acid, and aluminum. Environ Sci Technol 40:6015–6020
Cai L, Liu GH, Rensing C, Wang GJ (2009) Genes involved in arsenic transformation and resistance associated with different levels of arsenic-contaminated soils. BMC Microbiol 9:4
Challenger F (1945) Biological methylation. Chem Rev 36:315–361
Cullen WR, Reimer KJ (1989) Arsenic speciation in the environment. Chem Rev 89:713–764
Cummings DE, Caccavo F, Fendorf S, Rosenzweig RF (1999) Arsenic mobilization by the dissimilatory Fe(III)-reducing bacterium *Shewanella alga* BrY. Environ Sci Technol 33:723–729
Dittmar J, Voegelin A, Roberts LC, Hug SJ, Saha GC, Ali MA, Badruzzaman ABM, Kretzschmar R (2007) Spatial distribution and temporal variability of arsenic in irrigated rice fields in Bangladesh. 2. Paddy soil. Environ Sci Technol 41:5967–5972
Dittmar J, Voegelin A, Roberts LC, Hug SJ, Saha GC, Ali MA, Badruzzaman ABM, Kretzschmar R (2010) Arsenic accumulation in a paddy field in Bangladesh: seasonal dynamics and trends over a three-year monitoring period. Environ Sci Technol 44:2925–2931
Dixit S, Hering JG (2003) Comparison of arsenic(V) and arsenic(III) sorption onto iron oxide minerals: implications for arsenic mobility. Environ Sci Technol 37:4182–4189
Fendorf S, Eick MJ, Grossl P, Sparks DL (1997) Arsenate and chromate retention mechanisms on goethite. 1. Surface structure. Environ Sci Technol 31:315–320
Fendorf S, Herbel MJ, Tufano KJ, Kocar BD (2008) Biogeochemical processes controlling the cycling of arsenic in soils and sediments. In: Violante A, Huang PM, Gadd GM (eds) Biophysico-chemical processes of heavy metals and metalloids in soil environments. Wiley, Hoboken, pp 313–338
Gao S, Burau RG (1997) Environmental factors affecting rates of arsine evolution from and mineralization of arsenicals in soil. J Environ Qual 26:753–763
Garcia-Dominguez E, Mumford A, Rhine ED, Paschal A, Young LY (2008) Novel autotrophic arsenite-oxidizing bacteria isolated from soil and sediments. FEMS Microbiol Ecol 66: 401–410
Geiszinger A, Goessler W, Kosmus W (2002) Organoarsenic compounds in plants and soil on top of an ore vein. Appl Organomet Chem 16:245–249
Goldberg S, Johnston CT (2001) Mechanisms of arsenic adsorption on amorphous oxides evaluated using macroscopic measurements, vibrational spectroscopy, and surface complexation modeling. J Colloid Interface Sci 234:204–216
Grafe M, Eick MJ, Grossl PR (2001) Adsorption of arsenate(V) and arsenite(III) on goethite in the presence and absence of dissolved organic carbon. Soil Sci Soc Am J 65:1680–1687
Grafe M, Eick MJ, Grossl PR, Saunders AM (2002) Adsorption of arsenate and arsenite on ferrihydrite in the presence and absence of dissolved organic carbon. J Environ Qual 31:1115–1123
Grossl PR, Eick M, Sparks DL, Goldberg S, Ainsworth CC (1997) Arsenate and chromate retention mechanisms on goethite. 2. Kinetic evaluation using a pressure-jump relaxation technique. Environ Sci Technol 31:321–326

Hamon RE, Lombi E, Fortunati P, Nolan AL, McLaughlin MJ (2004) Coupling speciation and isotope dilution techniques to study arsenic mobilization in the environment. Environ Sci Technol 38:1794–1798

Hayakawa T, Kobayashi Y, Cui X, Hirano S (2005) A new metabolic pathway of arsenite: arsenic-glutathione complexes are substrates for human arsenic methyltransferase Cyt19. Arch Toxicol 79:183–191

Hossain MB, Jahiruddin M, Panaullah GM, Loeppert RH, Islam MR, Duxbury JM (2008) Spatial variability of arsenic concentration in soils and plants, and its relationship with iron, manganese and phosphorus. Environ Pollut 156:739–744

Hossain MB, Jahiruddin M, Loeppert RH, Panaullah GM, Islam MR, Duxbury JM (2009) The effects of iron plaque and phosphorus on yield and arsenic accumulation in rice. Plant Soil 317:167–176

Huang JH, Matzner E (2006) Dynamics of organic and inorganic arsenic in the solution phase of an acidic fen in Germany. Geochim Cosmochim Acta 70:2023–2033

Huang JH, Scherr F, Matzner E (2007) Demethylation of dimethylarsinic acid and arsenobetaine in different organic soils. Water Air Soil Pollut 182:31–41

Inskeep WP, McDermott TR, Fendorf S (2002) Arsenic (V)/(III) cycling in soils and natural waters: chemical and microbiological processes. In: Frankenberger JWT (ed) Environmental chemistry of arsenic. Marcel Dekker, New York, pp 183–215

Inskeep WP, Macur RE, Hamamura N, Warelow TP, Ward SA, Santini JM (2007) Detection, diversity and expression of aerobic bacterial arsenite oxidase genes. Environ Microbiol 9:934–943

Jain A, Loeppert RH (2000) Effect of competing anions on the adsorption of arsenate and arsenite by ferrihydrite. J Environ Qual 29:1422–1430

Jones CA, Langner HW, Anderson K, McDermott TR, Inskeep WP (2000) Rates of microbially mediated arsenate reduction and solubilization. Soil Sci Soc Am J 64:600–608

Khan MA, Islam MR, Panaullah GM, Duxbury JM, Jahiruddin M, Loeppert RH (2009) Fate of irrigation-water arsenic in rice soils of Bangladesh. Plant Soil 322:263–277

Khan MA, Stroud JL, Zhu YG, McGrath SP, Zhao FJ (2010) Arsenic bioavailability to rice is elevated in Bangladeshi paddy soils. Environ Sci Technol 44:8515–8521

Kirk G (2004) The biogeochemistry of submerged soils. Wiley, Chichester

Kocar BD, Fendorf S (2009) Thermodynamic constraints on reductive reactions influencing the biogeochemistry of arsenic in soils and sediments. Environ Sci Technol 43:4871–4877

Kocar BD, Herbel MJ, Tufano KJ, Fendorf S (2006) Contrasting effects of dissimilatory iron(III) and arsenic(V) reduction on arsenic retention and transport. Environ Sci Technol 40:6715–6721

Lafferty BJ, Loeppert RH (2005) Methyl arsenic adsorption and desorption behavior on iron oxides. Environ Sci Technol 39:2120–2127

Larsen S (1969) L value determination under paddy soil condition. Plant Soil 31:282–286

Li RY, Stroud JL, Ma JF, McGrath SP, Zhao FJ (2009) Mitigation of arsenic accumulation in rice with water management and silicon fertilization. Environ Sci Technol 43:3778–3783

Li HF, Lombi E, Stroud JL, McGrath SP, Zhao FJ (2010) Selenium speciation in soil and rice: influence of water management and Se fertilization. J Agric Food Chem 58:11837–11843

Lin S, Shi Q, Nix FB, Styblo M, Beck MA, Herbin-Davis KM, Hall LL, Simeonsson JB, Thomas DJ (2002) A novel S-adenosyl-L-methionine: arsenic(III) methyltransferase from rat liver cytosol. J Biol Chem 277:10795–10803

Liu GL, Cai Y (2010) Complexation of arsenite with dissolved organic matter. Conditional distribution coefficients and apparent stability constants. Chemosphere 81:890–896

Liu GL, Fernandez A, Cai Y (2011) Complexation of arsenite with humic acid in the presence of ferric iron. Environ Sci Technol 45:3210–3216

Lu Y, Adomako EE, Solaiman ARM, Islam MR, Deacon C, Williams PN, Rahman G, Meharg AA (2009) Baseline soil variation is a major factor in arsenic accumulation in Bengal delta paddy rice. Environ Sci Technol 43:1724–1729

Lumsdon DG, Fraser AR, Russell JD, Livesey NT (1984) New infrared band assignments for the arsenate ion adsorbed on synthetic goethite (alpha-FeOOH). J Soil Sci 35:381–386

Macur RE, Wheeler JT, McDermott TR, Inskeep WP (2001) Microbial populations associated with the reduction and enhanced mobilization of arsenic in mine tailings. Environ Sci Technol 35:3676–3682

Macur RE, Jackson CR, Botero LM, McDermott TR, Inskeep WP (2004) Bacterial populations associated with the oxidation and reduction of arsenic in an unsaturated soil. Environ Sci Technol 38:104–111

Maki T, Takeda N, Hasegawa H, Ueda K (2006) Isolation of monomethylarsonic acid mineralizing bacteria from arsenic contaminated soils of Ohkunoshima Island. Appl Organomet Chem 20:538–544

Manning BA, Goldberg S (1996) Modeling competitive adsorption of arsenate with phosphate and molybdate on oxide minerals. Soil Sci Soc Am J 60:121–131

Manning BA, Fendorf SE, Goldberg S (1998) Surface structures and stability of arsenic(III) on goethite: spectroscopic evidence for inner-sphere complexes. Environ Sci Technol 32:2383–2388

Marin AR, Masscheleyn PH, Patrick WH (1993) Soil redox-pH stability of arsenic species and its influence on arsenic uptake by rice. Plant Soil 152:245–253

Masscheleyn PH, Delaune RD, Patrick WH (1991) Effect of redox potential and pH on arsenic speciation and solubility in a contaminated soil. Environ Sci Technol 25:1414–1419

Meng XY, Qin J, Wang LH, Duan GL, Sun GX, Wu HL, Chu CC, Ling HQ, Rosen BP, Zhu YG (2011) Arsenic biotransformation and volatilization in transgenic rice. New Phytol 191:49–56

Mestrot A, Uroic MK, Plantevin T, Islam MR, Krupp EM, Feldmann J, Meharg AA (2009) Quantitative and qualitative trapping of arsines deployed to assess loss of volatile arsenic from paddy soil. Environ Sci Technol 43:8270–8275

Mestrot A, Feldmann J, Krupp EM, Hossain MS, Roman-Ross G, Meharg AA (2011) Field fluxes and speciation of arsines emanating from soils. Environ Sci Technol 45:1798–1804

Newman DK, Kennedy EK, Coates JD, Ahmann D, Ellis DJ, Lovley DR, Morel FMM (1997) Dissimilatory arsenate and sulfate reduction in *Desulfotomaculum auripigmentum* sp. nov. Arch Microbiol 168:380–388

Norra S, Berner ZA, Agarwala P, Wagner F, Chandrasekharam D, Stuben D (2005) Impact of irrigation with as rich groundwater on soil and crops: a geochemical case study in West Bengal delta plain, India. Appl Geochem 20:1890–1906

Ona-Nguema G, Morin G, Juillot F, Calas G, Brown GE (2005) EXAFS analysis of arsenite adsorption onto two-line ferrihydrite, hematite, goethite, and lepidocrocite. Environ Sci Technol 39:9147–9155

Onken BM, Hossner LR (1996) Determination of arsenic species in soil solution under flooded conditions. Soil Sci Soc Am J 60:1385–1392

Oscarson DW, Huang PM, Defosse C, Herbillon A (1981) Oxidative power of Mn(IV) and Fe(III) oxides with respect to As(IIII) in terrestrial and aquatic environments. Nature 291:50–51

Oscarson DW, Huang PM, Liaw WK, Hammer UT (1983) Kinetics of oxidation of arsenite by various manganese dioxides. Soil Sci Soc Am J 47:644–648

Panaullah GM, Alam T, Hossain MB, Loeppert RH, Lauren JG, Meisner CA, Ahmed ZU, Duxbury JM (2009) Arsenic toxicity to rice (*Oryza sativa* L.) in Bangladesh. Plant Soil 317:31–39

Peryea FJ (1991) Phosphate-induced release of arsenic from soils contaminated with lead arsenate. Soil Sci Soc Am J 55:1301–1306

Qin J, Rosen BP, Zhang Y, Wang GJ, Franke S, Rensing C (2006) Arsenic detoxification and evolution of trimethylarsine gas by a microbial arsenite S-adenosylmethionine methyltransferase. Proc Natl Acad Sci U S A 103:2075–2080

Radu T, Subacz JL, Phillippi JM, Barnett MO (2005) Effects of dissolved carbonate on arsenic adsorption and mobility. Environ Sci Technol 39:7875–7882

Raven KP, Jain A, Loeppert RH (1998) Arsenite and arsenate adsorption on ferrihydrite: kinetics, equilibrium, and adsorption envelopes. Environ Sci Technol 32:344–349

Reynolds JG, Naylor DV, Fendorf SE (1999) Arsenic sorption in phosphate-amended soils during flooding and subsequent aeration. Soil Sci Soc Am J 63:1149–1156

Ritter K, Aiken GR, Ranville JF, Bauer M, Macalady DL (2006) Evidence for the aquatic binding of arsenate by natural organic matter-suspended Fe(III). Environ Sci Technol 40:5380–5387

Roberts LC, Hug SJ, Dittmar J, Voegelin A, Saha GC, Ali MA, Badruzzaman ABM, Kretzschmar R (2007) Spatial distribution and temporal variability of arsenic in irrigated rice fields in Bangladesh. 1. Irrigation water. Environ Sci Technol 41:5960–5966

Roberts LC, Hug SJ, Dittmar J, Voegelin A, Kretzschmar R, Wehrli B, Cirpka OA, Saha GC, Ali MA, Badruzzaman ABM (2010) Arsenic release from paddy soils during monsoon flooding. Nat Geosci 3:53–59

Rochette EA, Bostick BC, Li GC, Fendorf S (2000) Kinetics of arsenate reduction by dissolved sulfide. Environ Sci Technol 34:4714–4720

Saha GC, Ali MA (2007) Dynamics of arsenic in agricultural soils irrigated with arsenic contaminated groundwater in Bangladesh. Sci Total Environ 379:180–189

Santini JM, Sly LI, Schnagl RD, Macy JM (2000) A new chemolithoautotrophic arsenite-oxidizing bacterium isolated from a gold mine: phylogenetic, physiological, and preliminary biochemical studies. Appl Environ Microbiol 66:92–97

Sharma P, Ofner J, Kappler A (2010) Formation of binary and ternary colloids and dissolved complexes of organic matter, Fe and As. Environ Sci Technol 44:4479–4485

Sharma P, Rolle M, Kocar B, Fendorf S, Kappler A (2011) Influence of natural organic matter on As transport and retention. Environ Sci Technol 45:546–553

Shimizu M, Arai Y, Sparks DL (2011a) Multiscale assessment of methylarsenic reactivity in soil. 1. Sorption and desorption on soils. Environ Sci Technol 45:4293–4299

Shimizu M, Arai Y, Sparks DL (2011b) Multiscale assessment of methylarsenic reactivity in soil. 2. Distribution and speciation in soil. Environ Sci Technol 45:4300–4306

Signes-Pastor A, Burlo F, Mitra K, Carbonell-Barrachina AA (2007) Arsenic biogeochemistry as affected by phosphorus fertilizer addition, redox potential and pH in a west Bengal (India) soil. Geoderma 137:504–510

Sparks DL (2003) Environmental soil chemistry. Academy Press, Amsterdam

Sposito G (1989) The chemistry of soils. Oxford University Press, New York

Stachowicz M, Hiemstra T, Van Riemsdijk WH (2007) Arsenic-bicarbonate interaction on goethite particles. Environ Sci Technol 41:5620–5625

Stroud JL, Khan MA, Norton GJ, Islam MR, Dasgupta T, Zhu YG, Price AH, Meharg AA, McGrath SP, Zhao FJ (2011a) Assessing the labile arsenic pool in contaminated paddy soils by isotopic dilution techniques and simple extractions. Environ Sci Technol 45:4262–4269

Stroud JL, Norton GJ, Islam MR, Dasgupta T, White R, Price AH, Meharg AA, McGrath SP, Zhao FJ (2011b) The dynamics of arsenic in four paddy fields in the Bengal delta. Environ Pollut 159:947–953

Sun XH, Doner HE (1996) An investigation of arsenate and arsenite bonding structures on goethite by FTIR. Soil Sci 161:865–872

Takahashi Y, Minamikawa R, Hattori KH, Kurishima K, Kihou N, Yuita K (2004) Arsenic behavior in paddy fields during the cycle of flooded and non-flooded periods. Environ Sci Technol 38:1038–1044

Takamatsu T, Aoki H, Yoshida T (1982) Determination of arsenate, arsenite, monomethylarsonate, and dimethylarsinate in soil polluted with arsenic. Soil Sci 133:239–246

Thomas DJ, Waters SB, Styblo M (2004) Elucidating the pathway for arsenic methylation. Toxicol Appl Pharmacol 198:319–326

Tufano KJ, Fendorf S (2008) Confounding impacts of iron reduction on arsenic retention. Environ Sci Technol 42:4777–4783

Tufano KJ, Reyes C, Saltikov CW, Fendorf S (2008) Reductive processes controlling arsenic retention: revealing the relative importance of iron and arsenic reduction. Environ Sci Technol 42:8283–8289

Ventura W, Watanabe I, Castillo MB, Delacruz A (1981) Involvement of nematodes in the soil sickness of a dryland rice-based cropping system. Soil Sci Plant Nutr 27:305–315

Violante A, Pigna M (2002) Competitive sorption of arsenate and phosphate on different clay minerals and soils. Soil Sci Soc Am J 66:1788–1796

Waltham CA, Eick MJ (2002) Kinetics of arsenic adsorption on goethite in the presence of sorbed silicic acid. Soil Sci Soc Am J 66:818–825

Warwick P, Inam E, Evans N (2005) Arsenic's interaction with humic acid. Environ Chem 2:119–124

Wenzel WW, Brandstetter A, Wutte H, Lombi E, Prohaska T, Stingeder G, Adriano DC (2002) Arsenic in field-collected soil solutions and extracts of contaminated soils and its implication to soil standards. J Plant Nutr Soil Sci 165:221–228

Williams PN, Zhang H, Davison W, Meharg AA, Hossain M, Norton GJ, Brammer H, Islam MR (2011) Organic matter-solid phase interactions are critical for predicting arsenic release and plant uptake in Bangladesh paddy soils. Environ Sci Technol 45:6080–6087

Woolson EA, Kearney PC (1973) Persistence and reactions of ^{14}C-cacodylic acid in soils. Environ Sci Technol 7:47–50

Woolson EA, Aharonson N, Iadevaia R (1982) Application of the high-performance liquid-chromatography flameless atomic-absorption method to the study of alkyl arsenical herbicide metabolism in soil. J Agric Food Chem 30:580–584

Xu H, Allard B, Grimvall A (1988) Influence of pH and organic substance on the adsorption of As(V) on geologic materials. Water Air Soil Pollut 40:293–305

Xu H, Allard B, Grimvall A (1991) Effects of acidification and natural organic materials on the mobility of arsenic in the environment. Water Air Soil Pollut 57–8:269–278

Xu XY, McGrath SP, Meharg A, Zhao FJ (2008) Growing rice aerobically markedly decreases arsenic accumulation. Environ Sci Technol 42:5574–5579

Yamaguchi N, Nakamura T, Dong D, Takahashi Y, Amachi S, Makino T (2011) Arsenic release from flooded paddy soils is influenced by speciation, Eh, pH, and iron dissolution. Chemosphere 83:925–932

Yang JK, Barnett MO, Jardine PM, Basta NT, Casteel SW (2002) Adsorption, sequestration, and bioaccessibility of As(V) in soils. Environ Sci Technol 36:4562–4569

Yoshinaga M, Cai Y, Rosen BP (2011) Demethylation of methylarsonic acid by a microbial community. Environ Microbiol 13:1205–1215

Yuan CG, Lu XF, Qin J, Rosen BP, Le XC (2008) Volatile arsenic species released from *Escherichia coli* expressing the AsIII S-adenosylmethionine methyltransferase gene. Environ Sci Technol 42:3201–3206

Zhang H, Selim HM (2008) Reaction and transport of arsenic in soils: equilibrium and kinetic modeling. Adv Agron 98:45–115

Zhao FJ, McGrath SP, Meharg AA (2010) Arsenic as a food-chain contaminant: mechanisms of plant uptake and metabolism and mitigation strategies. Ann Rev Plant Biol 61:535–559

Chapter 6
The Physiology of Arsenic in Rice

6.1 Introduction

As discussed in Chap. 5, arsenic may be present in different chemical forms in paddy soil; the most common forms are inorganic arsenic (arsenate and arsenite) and methylated species such as MMA and DMA. Plant roots are able to absorb these arsenic species from the soil solution; however, the mechanisms involved and the rates of uptake differ greatly between arsenic species. In this chapter, we will discuss the mechanisms of arsenic uptake, translocation and metabolism, focusing on rice but also draw on the knowledge gained from studies of other plant species. Accumulation of excess arsenic can also lead to phytotoxicity and yield losses. The "straight-head disease", a common physiological disorder in paddy rice, is very likely caused by the toxicity of methylated arsenic species; this is also discussed in the present chapter.

6.2 Mechanisms of Arsenic Uptake by Roots

It is important to consider protonation/dissociation of arsenic species as influenced by pH, because these processes determine the charge characteristics of the molecules, which have a strong bearing on their transport across the biological membranes of plant cells. Arsenate has low dissociation constants, with pK_{a1} and pK_{a2} being 2.2 and 6.97, respectively. Under the normal pH range encountered in paddy soils or inside plant cells (typically 5.0 to 8.0), arsenate is dissociated and present predominantly as $H_2AsO_4^-$ and $HAsO_4^{2-}$ (Fig. 6.1). In contrast, arsenite is mostly undissociated and present as neutral molecules of arsenous acid (H_3AsO_3) because of its high pK_{a1} (9.2). Both the neutral and dissociated forms of DMA (pK_{a1} 6.1) and MMA (pK_{a1} 4.2) are present, although for MMA more will be dissociated (Fig. 6.1).

A.A. Meharg and F.J. Zhao, *Arsenic & Rice*, DOI 10.1007/978-94-007-2947-6_6,

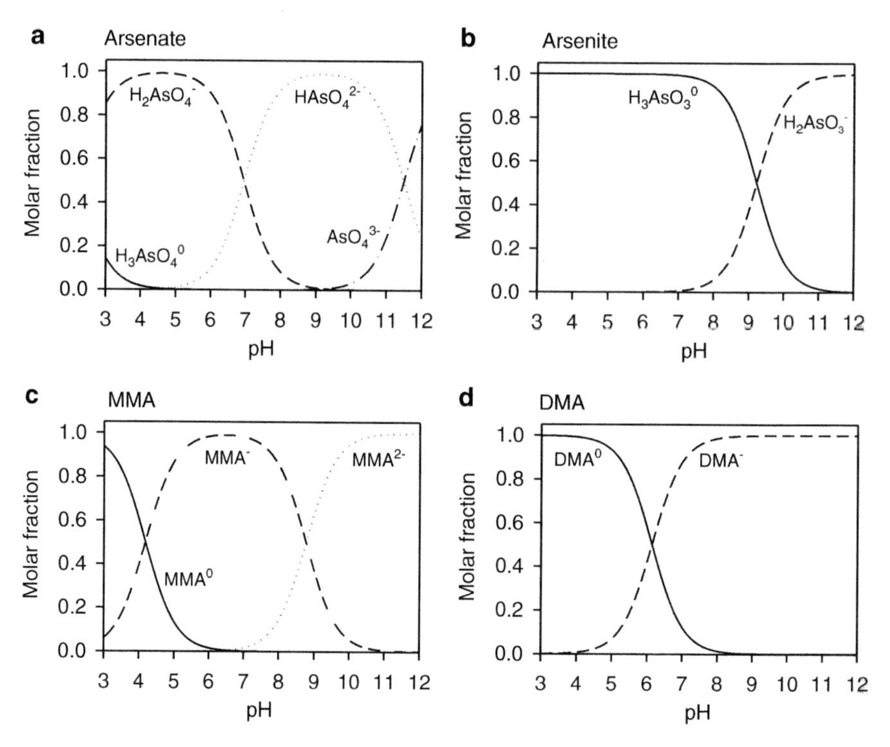

Fig. 6.1 Dissociation of arsenate, arsenite, DMA and MMA as influenced by pH

6.2.1 Uptake of Arsenate

Arsenate is a chemical analogue of phosphate because the two elements belong to the same group (Group VB) in the Periodic Table. For this reason, the plant membrane transporters for phosphate cannot discriminate between phosphate and arsenate and, as a result, arsenate is taken up into root cells via these transporters. There have been many physiological studies showing that phosphate effectively suppresses arsenate uptake in hydroponic cultures, a result that can be attributed to the competition between the two anions for the same transporter(s); examples include barley (Asher and Reay 1979), *Holcus lanatus* (Meharg and Macnair 1992) and rice (Abedin et al. 2002). Conversely, plants starved of phosphorus have enhanced abilities for the uptake of both phosphate and arsenate (Lee 1982; Wang et al. 2002); this is because P-starvation up-regulates the expression of phosphate transporter genes leading to more transporter proteins on the membranes (Raghothama 1999). Direct evidence for arsenate sharing the phosphate uptake pathway comes from the studies of phosphate transport mutants. When genes involved in phosphate uptake are mutated, the mutant plants become more tolerant to arsenate because its uptake is restricted (González et al. 2005; Shin et al. 2004; Wu et al. 2011). In fact, the arsenate tolerance test can be used as a rapid screening method to isolate phosphate transport mutants.

The affinity of a substrate to a transporter can be gauged from the Michaelis-Menten kinetic parameter K_m, which, together with the maximum influx velocity (V_{max}), can be measured by short-term uptake kinetics. V_{max} is a measure of the abundance or activity of the transporter protein in the plasma membrane. Phosphate transporters can be divided into two types; those operating in the low concentration (μM) range are called high-affinity transporters (with low K_m) and those in the high concentration (mM) range low-affinity transporters (with high K_m). For root uptake of phosphate, K_m values typically vary from 2 to 10 μM, indicating that high-affinity transporters are in operation (Barber 1984; Raghothama 1999). Phosphate is usually present at very low concentrations in the soil solution (typically <10 μM); the concentration near the root surface may be even lower because the diffusion of phosphate ions from the bulk soil to the root surface is usually slower than the rate of root absorption, thus causing a depletion in the rhizosphere (Barber 1984). This means that plant roots must employ high-affinity transporters to scavenge low levels of phosphate in the soil solution. Low-affinity transporters are unlikely to play a significant role in phosphate uptake from the soil solution, but may be involved in the loading of phosphate into xylem vessels.

For arsenate, Abedin et al. (2002) reported K_m of 1.8–14 μM (mean 6.1 μM) in eight cultivars of rice commonly grown in Bangladesh. These values are comparable to those reported for phosphate. Wu et al. (2011) compared arsenate and phosphate uptake kinetics in the rice cultivar Nipponbare and obtained similar K_m values for the two anions; in contrast, the V_{max} for phosphate was two- to three-fold higher than that for arsenate. These results mean that phosphate is taken up faster than arsenate, even though the affinities are similar. In the same study, arsenate uptake was inhibited greatly by the presence of phosphate in the assay medium. Moreover, the kinetics of arsenate uptake changed from hyperbolic (i.e. the Michaelis-Menten kinetics) to linear in the presence of phosphate, indicating a marked loss of affinity for arsenate due to the competition of phosphate (Fig. 6.2). In the rice plants over-expressing the phosphate transporter gene *OsPht1;8* (sometimes also abbreviated as *OsPT8*; the prefix *Os* stands for *Oryza sativa*), arsenate uptake was much enhanced, resulting in a 5.5 fold increase in the V_{max} but also a doubling of the K_m (Fig. 6.2) (Wu et al. 2011).

There are 13 genes encoding the putative high-affinity phosphate transporter proteins in the rice genome (Paszkowski et al. 2002). These are named *OsPht1;1* – *OsPht1;13* (Note: italic is used to denote genes while non-italic for proteins). The word "putative" is used here because the affinity has not been determined for most of these transporters. These genes are expressed in different tissues and the proteins they encode may play different roles of phosphate uptake, cellular P homeostasis and long-distance transport. For example, OsPht1;6 is localized in both epidermal and cortical cells of rice roots and appears to play a broad role in phosphate uptake and translocation (Ai et al. 2009). In contrast, OsPht1;2 is localized exclusively in the stele of rice roots, probably being involved in the loading of phosphate into the xylem vessels (Ai et al. 2009). Heterologous functional assays in *Xenopus laevis*

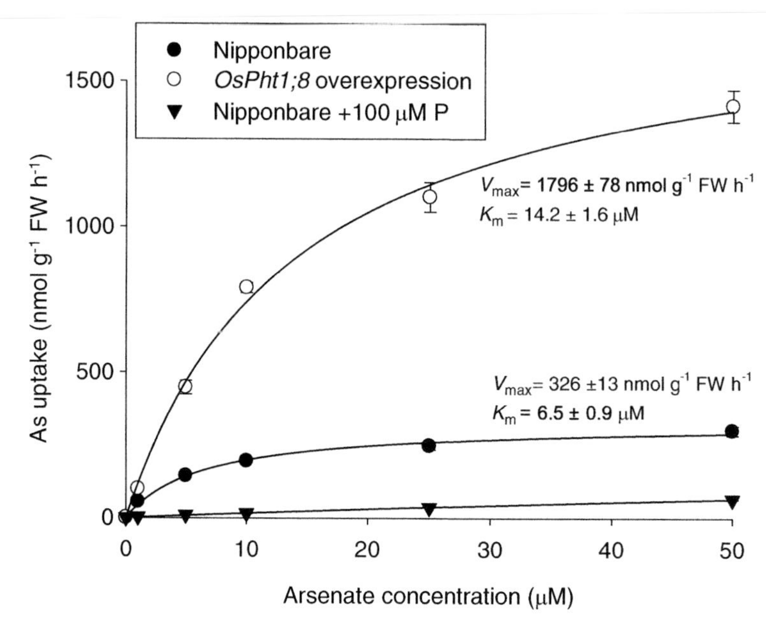

Fig. 6.2 Arsenate uptake kinetics of the rice cultivar Nipponbare with or without the presence of 100 μM phosphate, and of the *OsPht1;8* over-expressing line (Redrawn with permission from Wu et al. (2011). Copyright American Society of Plant Biologists)

oocytes showed that OsPht1;2 has a relatively low affinity for phosphate (Ai et al. 2009). *OsPht1;8* is expressed in the root epidermal cells, as well as in the root-shoot junction and leaves (Jia et al. 2011). OsPht1;8 has a high affinity for both phosphate and arsenate and is probably involved in the uptake and long-distance transport of phosphate and arsenate in rice (Jia et al. 2011; Wu et al. 2011). The gene *OsPht1;11* was found to be specifically induced during the arbuscular mycorrhizal symbiosis of rice roots grown under aerobic conditions, suggesting a role in phosphate acquisition in the root-mycorrhizal symbiont (Paszkowski et al. 2002). It would be of great interest to understand how different phosphate transporters in rice contribute to arsenate uptake and transport within plants, and whether they differ in the relative affinities for phosphate and arsenate.

Because phosphate and arsenate are anions, their uptake into root cells is against the electrochemical gradient and is driven by the proton motive force generated by a plasma membrane H^+ ATPase. Uptake of phosphate (Raghothama 1999) and arsenate (Ullrich-Eberius et al. 1989) is accompanied by a co-transport of 2–4 H^+ for each molecule of the anions transported. A decrease in the external medium pH may enhance arsenate uptake as it does phosphate uptake (Sentenac and Grignon 1985). In the soil-plant system, however, the pH effect is likely to be more complex because pH also affects arsenate adsorption in soil (Zhang and Selim 2008) (see also Chap. 5).

6.2.2 Uptake of Arsenite

An important biogeochemical feature of paddy soil is the prevailing anaerobic conditions developed after submergence with water; such conditions are conducive to the reduction of arsenate to arsenite (see Chap. 5). In the soil solution of flooded paddy soils, arsenite becomes the predominant arsenic species (Panaullah et al. 2009; Stroud et al. 2011; Takahashi et al. 2004; Xu et al. 2008). Therefore, uptake of arsenite assumes great importance for paddy rice.

Uptake kinetic studies with rice have shown that the maximum influx velocity (V_{max}) for arsenite is comparable to that for arsenate (Abedin et al. 2002). However, K_m for arsenite (6–52 μM, mean = 19 μM) is larger than that for arsenate, indicating a lower affinity. Based on the inhibitive effect of glycerol on arsenite influx into rice roots, and the fact that arsenite is transported into *Escherichia coli*, yeast and mammalian cells by aquaglyceroporin channels, Meharg and Jardine (2003) suggested a similar mechanism for arsenite uptake by plant roots. This has been confirmed by recent molecular studies which have identified a number of plant Nodulin 26-like Intrinsic Proteins (NIPs) as permeable to arsenite (Bienert et al. 2008b; Isayenkov and Maathuis 2008; Kamiya et al. 2009; Ma et al. 2008). NIPs are a subfamily of plant aquaporins permeable to a wide range of neutral substrates, including glycerol, urea, ammonia, silicic acid and boric acid (Maurel et al. 2008). It is not surprising that arsenite can be transported through NIP aquaporins because it is mostly undissociated under normal environmental and physiological pH (Fig. 6.1). NIPs are also called plant aquaglyceroporins (Wallace et al. 2006), although some of the NIPs (e.g. OsNIP2;1) have little permeability to glycerol (Ma et al. 2006; Mitani et al. 2008), and there is no direct evidence for a physiological role in plants of glycerol transport through NIPs (Bienert et al. 2008a). In comparison with the Plasma membrane Intrinsic Proteins (PIPs) whose main functions are water transport, NIPs have larger pore sizes thus allowing the passage of neutral molecules larger than water, but have only low or even no water permeability (Bansal and Sankararamakrishnan 2007; Wallace et al. 2006). Two other subfamilies of plant aquaporins are Tonoplast Intrinsic Proteins (TIPs) and Small basic Intrinsic Proteins (SIPs); whether they are permeable to arsenite remains unknown.

There are 9 NIP genes in the Arabidopsis genome and 13 members in the rice genome (Forrest and Bhave 2007; Maurel et al. 2008). All NIPs tested so far appear to be permeable to arsenite in heterologous assays using yeast or *Xenopus* oocytes; in contrast, the permeability for silicic acid and boric acid is restricted to only a few NIP members (reviewed in Zhao et al. 2009; Zhao et al. 2010b). The selectivity of NIP proteins for arsenite is relatively low compared with that for the substrates silicic acid and boric acid (Mitani-Ueno et al. 2011). An *in planta* role of arsenite transport has been established for the Arabidopsis AtNIP1;1 and AtNIP7;1 (Isayenkov and Maathuis 2008; Kamiya et al. 2009) and for rice OsNIP2;1 (Ma et al. 2008). Because arsenic is not essential for plants, its uptake by plants is adventitious via transporters for nutrients or substances that are important for plant growth. As discussed below, the primary function of OsNIP2;1 is to transport silicic acid. For AtNIP1;1 and AtNIP7;1, the primary functions are not yet known.

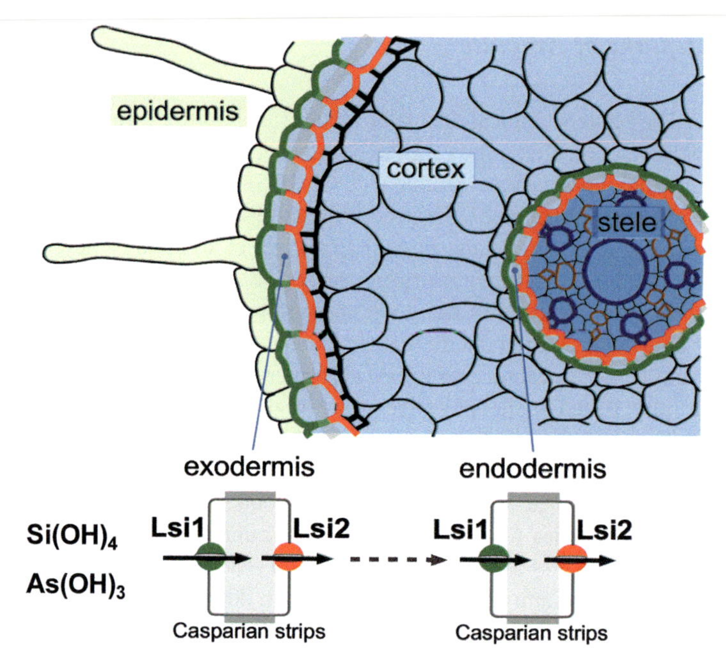

Fig. 6.3 Arsenite uptake through the silicon transport pathway in rice roots. *Green* and *red colours* indicate the localisation of Lsi1 and Lsi2 proteins, respectively (Redrawn with permission from Zhao et al. (2009))

OsNIP2;1 (also named Lsi1) is a major influx transporter for silicic acid in rice (Ma et al. 2006). The *Lsi1* gene is constitutively highly expressed in rice roots, especially in the mature zone. The protein is localized to the distal side of the plasma membranes in the exodermal and endoderma cells of rice roots, mediating the entry of silicic acid into the cells (Fig. 6.3). Mutation in this gene results in a marked decrease in Si uptake by rice (Ma et al. 2006). Silicic acid and arsenous acid have similarly high pK_a values (around 9.2) and also similar molecular size [$Si(OH)_4$ being slightly larger than $As(OH)_3$)]. The possibility of Lsi1 being involved in arsenite uptake into rice roots was tested by Ma et al. (2008). After exposure to 2 μM arsenite for 1 day, the rice *lsi1* mutant took up a significantly smaller amount of arsenite than the wild-type plants (As concentrations in roots and shoots being only about 50% and 30%, respectively, of the wild-type; Fig. 6.4). Furthermore, when the *Lsi1* gene was expressed in *Xenopus* oocytes, uptake of arsenite was greatly enhanced. Three of the other known NIP proteins from rice (OsNIP1;1, OsNIP2;2 and OsNIP3;1) are also able to mediate arsenite transport into *Xenopus* oocytes to varying degrees; however, their expression levels in rice roots are very low and therefore unlikely to play a significant role in arsenite uptake (Ma et al. 2008).

In long-term experiments (e.g. rice plants grown to maturity), Lsi1 mutation had rather little effect on arsenic accumulation in the shoots or grain (Ma et al. 2008). A possible explanation is that aquaporin channels like Lsi1 allow bi-directional transport depending on the concentration gradient of the substrate (Zhao et al. 2010a).

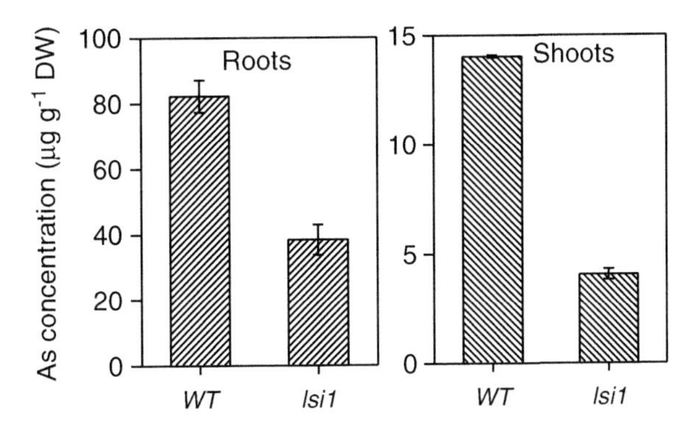

Fig. 6.4 Effect of Lsi1 (OsNIP2;1) mutation on arsenite uptake by rice after exposure to 2 μM arsenite for 1 day (mean ± SE); *WT* wild-type, *lsi1* Lsi1 mutant (Redrawn with permission from Ma et al. (2008))

Furthermore, the translocation of arsenic from roots to shoots is the bottleneck of the arsenic accumulation in the above-ground tissues (Zhao et al. 2009), whereas arsenic entry into the root cells may be a less decisive factor.

A second type of Si transporter is also involved in arsenite transport. This transporter, named Lsi2, is an efflux carrier localized to the proximal side of the plasma membranes in the exodermal and endodermal cells of rice roots, i.e. in the same cells, but the opposite side, where Lsi1 is located (Fig. 6.3) (Ma et al. 2007). Lsi2 transports Si, possibly silicic acid, out of the exodermal and endodermal cells toward the stele and, unlike Lsi1, this process appears to be active. Lsi2 mutation also results in a dramatic decrease in Si accumulation in rice shoots. When *lsi2* mutant and its wild-type rice plants were exposed to arsenite for 1 day, there was little difference in the concentration of arsenic in the roots, not surprising because Lsi2 is not an influx transporter. However, arsenite concentration in the xylem sap and the total arsenic concentration in the shoots of the mutant were decreased by more than 90% (Fig. 6.5) (Ma et al. 2008). In the wild-type plants, an addition of 0.5 mM silicic acid to the nutrient solution decreased arsenite concentration in the xylem sap (by 40%) and the total arsenic concentration in the shoots (by 33%); this effect was not seen in the *lsi2* mutant because the transporter was not functional. Compared with Lsi1 mutation, Lsi2 mutation produced a much larger effect on arsenic accumulation in the above-ground tissues including grain in field-grown plants; straw and grain arsenic concentrations were decreased by >80% and >40%, respectively. These results establish Lsi2 as a key arsenite transporter in rice. In fact, Lsi2 has a low level of homology with the *E. coli* arsenite efflux transporter ArsB.

The polar distribution of Lsi1 and Lsi2 appears to be unique to rice; in other cereals (e.g. barley and maize) the efflux carrier Lsi2 is localized on all sides of the plasma membranes of the endodermal cells (Mitani et al. 2009). This polar localization pattern allows rice to take up Si efficiently, as Si plays important roles in the resistance against biotic and abiotic stresses (Ma and Yamaji 2006). Typically, rice accumulates 5–10% Si in the shoots. The efficient Si uptake pathway also allows the inadvertent uptake of arsenite. In hydroponic culture, rice transports a significantly larger proportion of arsenic to the above-ground tissue than wheat or barley (Su et al. 2010). When combined with a massive arsenite mobilization induced by

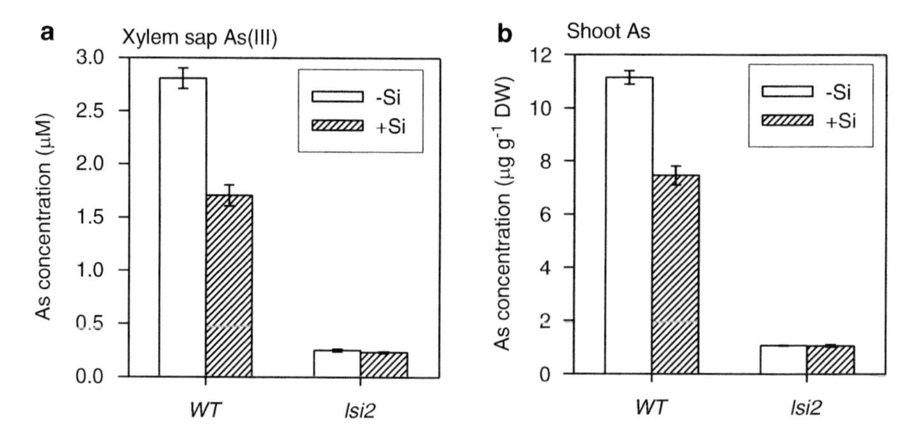

Fig. 6.5 Effect of Lsi2 mutation on arsenite concentration in xylem sap and arsenic accumulation in rice shoots. Plants were exposed to 5 μM arsenite with or without 0.5 mM silicic acid for 1 day (mean ± SE); *WT* wild-type, *lsi2* Lsi2 mutant (Redrawn with permission from Ma et al. (2008))

the anaerobic conditions in submerged paddy soil, it is not surprising that rice is disconcertingly efficient at arsenic accumulation (Williams et al. 2007b).

Evidence for arsenite and Si sharing the same transport pathway in rice also comes from the observation that arsenic concentrations in straw and grain correlate negatively with available Si in soils (Fig. 6.6) (Bogdan and Schenk 2008). Moreover, when sparingly soluble silicon gel (SiO_2) was added to soil in a pot experiment, arsenic accumulation in straw and grain was decreased markedly (Li et al. 2009b).

It should be pointed out that the competitive effect of silicic acid on arsenite uptake occurs during the Lsi2 transport, rather than during the Lsi1 transport. Experiments with *Xenopus* oocytes expressing the *Lsi1* gene showed no inhibition of arsenite transport by silicic acid up to 0.5 mM. This lack of inhibition is thought to be due to the extremely fast flux of solutes through the channel protein, thus masking the effect of competition. In contrast, Lsi2 is not a channel, but a carrier protein, for which competitions between transport substrates are expected. It has been reported that addition of silicic acid to the nutrient solution inhibited arsenic accumulation by rice exposed to arsenate, through an unknown effect not due to a direct competition between Si and arsenate (Guo et al. 2007). This effect can be explained by a rapid reduction of arsenate to arsenite in roots (Xu et al. 2007), followed by arsenite transport via Lsi2 which is subject to Si inhibition.

6.2.3 *Uptake of Methylated Arsenic Species*

It has become clear recently that plants lack the ability to methylate arsenic (see Sect. 6.3.3). It thus follows that methylated arsenic species detected in rice plants and grain must have originated from the soil (see Chap. 5). Uptake of the methylated arsenic species MMA and DMA has been studied in a number of plant

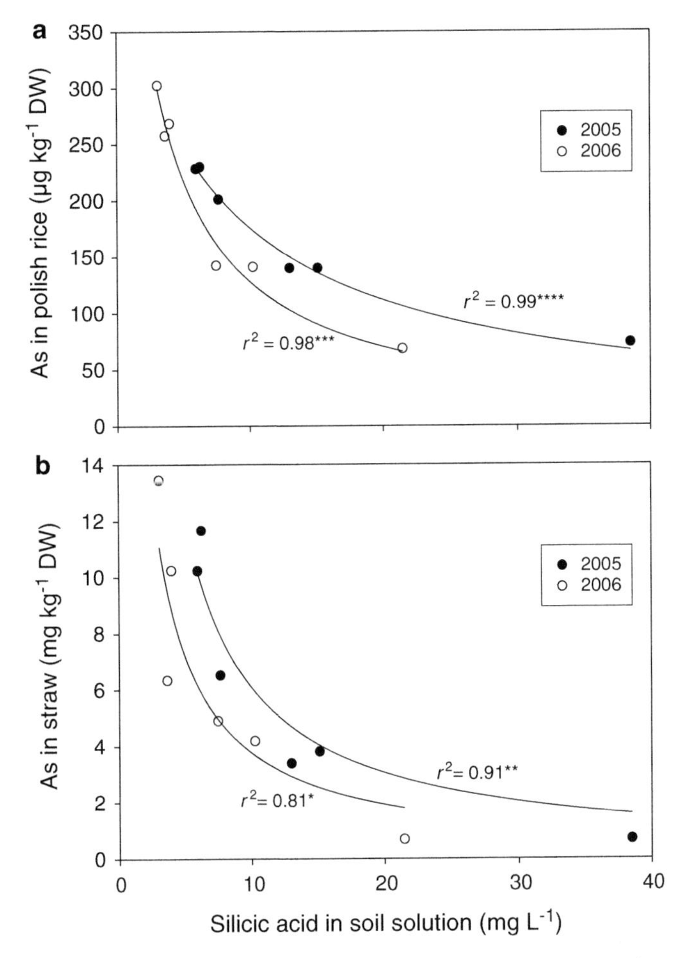

Fig. 6.6 Relationship between silicic acid concentration in soil solution and arsenic concentration in polished rice grain (**a**) and straw (**b**) sampled in 2005 and 2006 (Redrawn with permission from Bogdan and Schenk (2008). Copyright (2008) American Chemical Society)

species in hydroponic culture. In general, they are taken up by roots more slowly than inorganic arsenic, but then they are transported in xylem and phloem more efficiently (see Sect. 6.5) (Abedin et al. 2002; Carbonell-Barrachina et al. 1998, 1999; Carey et al. 2010; Marin et al. 1992; Raab et al. 2007b; Ye et al. 2010).

Uptake kinetics of MMA by rice roots appear to be saturable (i.e. conforming to the Michalis-Menten kinetics), whereas those of DMA are linear (Abedin et al. 2002; Li et al. 2009a). Uptake of both arsenic species, particularly MMA, is sensitive to the pH of the medium (Fig. 6.7) (Li et al. 2009a). Uptake decreases with increasing pH from 4.5 to 6.5, and this trend is broadly in line with the decreasing percentage of the undissociated molecules of MMA and DMA, suggesting that the undissociated molecules are the main species taken up by rice roots. This pH dependency also

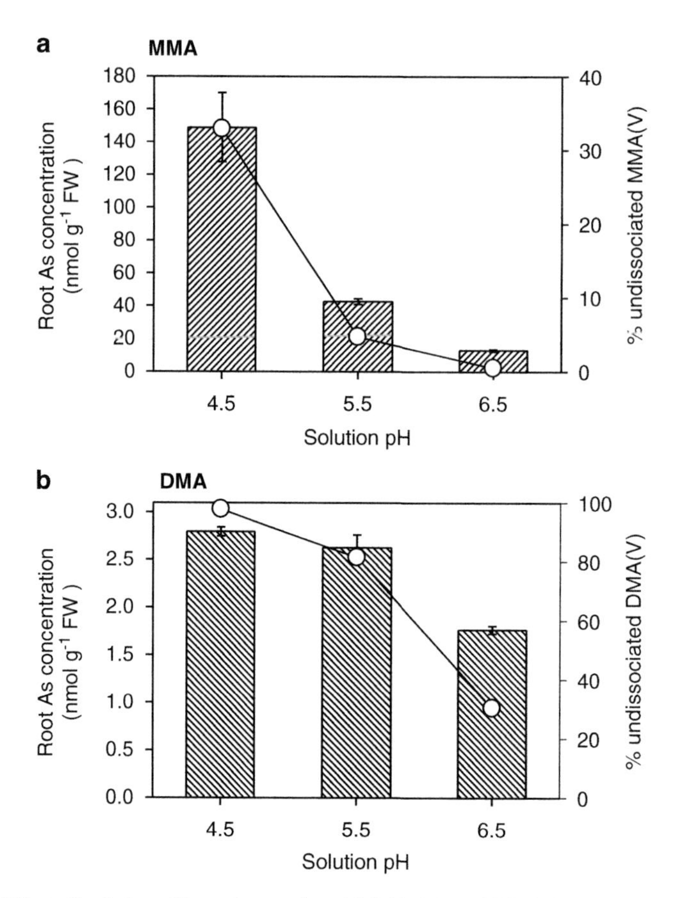

Fig. 6.7 Effect of solution pH on the uptake of MMA (**a**) and DMA (**b**) by rice. Plants were exposed to 5 μM MMA or DMA for 24 h. *Bars* represent arsenic concentrations in roots and *open circles and lines* the % of undissociated MMA or DMA. The trend is similar for the arsenic concentrations in xylem sap and in shoots (Redrawn with permission from Li et al. (2009a). Copyright American Society of Plant Biologists)

explains the inconsistency in previous studies comparing uptake of methylated arsenic versus inorganic arsenic. For example, Marin et al. (1992) reported efficient uptake of MMA by rice, most likely because they used a low pH in the nutrient solution (pH 4.0), whereas others (Abedin et al. 2002; Raab et al. 2007b) observed low uptake efficiency of MMA by rice using media with higher pHs.

Li et al. (2009a) presented evidence for the uptake of undissociated MMA and DMA through the rice aquaporin Lsi1 (OsNIP2;1). Compared with the wild-type rice, the rice *lsi1* mutant lost about 80% and 50% of the capacity to take up MMA and DMA, respectively. Expression of the *Lsi1* gene in *Xenopus* oocytes also significantly increased the uptake of MMA. In the case of DMA, its uptake was too low to be reliably tested in the oocyte assay. In the grain samples harvested from a paddy field, the *lsi1* mutants also contained a significantly lower concentration of DMA than the wild-type rice (Zhao and Ma, unpublished data); MMA was not detectable in both samples as are most field-grown samples. Therefore, Lsi1 is permeable not only to silicic acid and arsenite, but also to neutral molecules of MMA and DMA. Interestingly, the Lsi2 mutation had no effect on the uptake and root-to-shoot translocation of MMA and DMA (Li et al. 2009a). Once taken inside the root cells, both MMA and DMA will dissociate at the cytoplasmic pH (~7.5), and the negatively charged molecules may be transported via other mechanisms.

Although uptake of undissociated MMA and DMA via Lsi1 is an important mechanism, it is possible that some dissociated MMA and DMA are also taken up, albeit at a lower rate than the neutral molecules. Figure 6.7 shows that the uptake of MMA at pH 5.5 and 6.5, and of DMA at pH 6.5, is greater than that predicted by the percentage of the undissociated molecules. This observation may be attributed to additional uptake of dissociated molecules. The mechanism for the uptake of dissociated MMA and DMA is not yet known. Abbas and Meharg (2008) reported that DMA uptake by maize seedlings was enhanced by phosphorus deficiency, suggesting a possible involvement of phosphate transporters in the uptake of dissociated DMA.

6.3 Transformation of Arsenic Species in Plants

6.3.1 Arsenate Reduction

Most plant species tested to date, including rice, are able to reduce arsenate to arsenite rapidly (Zhao et al. 2009). As(III) is often the dominant oxidation state of arsenic in plants exposed to arsenate (e.g. Dhankher et al. 2002; Pickering et al. 2000; Xu et al. 2007). For example, after exposure to 10 μM arsenate for 1 day in hydroponic culture, rice roots contained 92% As(III) and 8% As(V) (Xu et al. 2007). In general, the proportion of As(III) in shoots is larger than that in roots, suggesting that both roots and shoots are able to reduce arsenate.

Based on the sequence homology with the yeast arsenate reductase Acr2p, plant homologues of ACR2 have been cloned from *Arabidopsis thaliana* (Dhankher et al. 2006), *Holcus lanatus* (Bleeker et al. 2006), *Pteris vittata* (Ellis et al. 2006) and rice (Duan et al. 2007). Heterologous expression of the rice *OsACR2;1* gene in the *E. coli* and yeast mutants lacking the indigenous arsenate reductase restored their resistance to arsenate, and purified OsACR2 protein from *E. coli* overexpressing *OsACR2;1* can catalyze arsenate reduction using reduced glutathione as a reductant (Duan et al. 2007). In comparison, the protein encoded by the second gene *OsACR2;2* appears to be much less effective. *OsACR2;1* was found to be expressed in both roots and shoots of rice with the expression being enhanced by arsenate exposure (Duan et al. 2007). Yeast Acr2p and plant ACR2 proteins are members of the protein tyrosine phosphatase (PTPase) superfamily and their ability to reduce arsenate may be a secondary function.

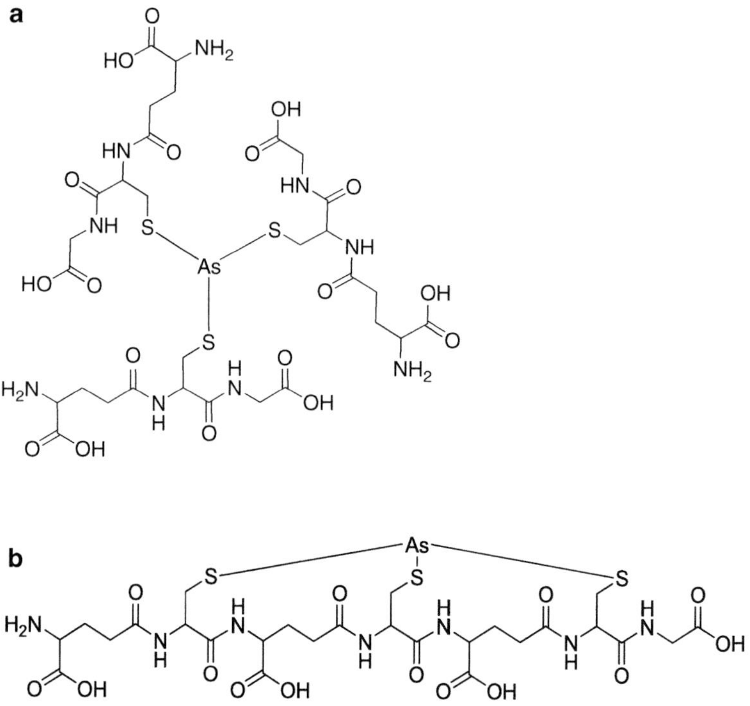

Fig. 6.8 Structures of As(III)-GS_3 (**a**) and As(III)-PC_3 (**b**)

In addition to ACR2, there are likely to be other plant enzymes capable of reducing arsenate (Bleeker et al. 2006). This is demonstrated by the lack of an effect on arsenate reduction in the Arabidopsis mutants lacking AtACR2 (Zhao et al. 2009). It is also possible that arsenate may be reduced non-enzymatically by reaction with glutathione inside plant cells, although the significance of this reduction remains unknown.

Reduction of arsenate to arsenite diverts arsenic away from the phosphate transport and metabolism pathways. This process allows plants to detoxify arsenic more easily through complexation.

6.3.2 Arsenite Complexation

Arsenite has a high affinity for the thiol (–SH) groups of peptides or proteins. Figure 6.8a shows a complex of arsenite bound by three – SH groups from the cysteinyl residues of three molecules of the tripeptide glutathione (GSH), which has been shown to form *in vitro* (Delnomdedieu et al. 1994). Proteins with vicinal thiol groups appear to be particularly sensitive to arsenite binding, which leads to structural alteration or inactivation of the catalytic function of the enzymes (Hughes 2002). In *Arabidopsis thaliana*, the plastidial lipoamide dehydrogenase of the pyruvate dehydrogenase complex is a sensitive target of arsenite binding and toxicity

(Chen et al. 2010). To detoxify arsenite, plants synthesize phytochelatins (PCs) from GSH. PCs have the general structure of $(\gamma\text{-GluCys})_n$-Gly, where n is generally in the range of 2–5 (but can be up to 11) (Cobbett and Goldsbrough 2002). Other variants of phytochelatins also exist with the terminal glycine residue being replaced by other amino acids (e.g. alanine, serine, glutamate). The cysteine residues of PCs provide –SH groups to complex arsenite, as shown in Fig. 6.8b for the As(III)-PC_3 complex. Exposure to arsenate or arsenite enhances PC synthesis markedly (e.g. Schmöger et al. 2000; Sneller et al. 1999). Both the PC-deficient and GSH-deficient mutants of *Arabidopsis thaliana* are hypersensitive to arsenic (Ha et al. 1999; Liu et al. 2010); the latter also produces very little PCs because of a shortage of the substrate GSH (Howden et al. 1995). In contrast, enhancing the PC biosynthesis capacity by over-expressing relevant genes has the effect of increasing arsenic tolerance in plants (e.g. Dhankher et al. 2002; Guo et al. 2008; Li et al. 2004). It is clear that PCs play an important role in arsenic detoxification in plants; an exception is the arsenic hyperaccumulator *Pteris* species which do not appear to rely on PCs for arsenic detoxification (Zhao et al. 2009).

A number of As(III)-thiol complexes have been identified in plants, including As(III)-$(PC_2)_2$, As(III)-PC_3, GS-As(III)-PC_2 and As(III)-PC_4 (Liu et al. 2010; Raab et al. 2004a, 2005, 2007a). Trivalent MMA(III) can also be complexed by PCs (Raab et al. 2005). Exposure of rice seedlings to arsenate upregulates the expression of a number of genes (Norton et al. 2008) or enzymes (Ahsan et al. 2008) involved in glutathione synthesis, metabolism, and transport, supporting the role of GSH and PCs in arsenic detoxification in rice.

6.3.3 Arsenic Methylation

Methylated arsenic species, especially DMA, accounts for a considerable proportion of the total arsenic in rice grain (see Chap. 2). Methylated arsenic species are also found in plant samples collected from their natural habitats (Geiszinger et al. 2002; Kuehnelt et al. 2000) or from non-sterile hydroponic or sand cultures (Mihucz et al. 2005; Nissen and Benson 1982; Quaghebeur and Rengel 2003; Raab et al. 2007a). It was thought that plants were able to methylate inorganic arsenic, particularly under the conditions of phosphorus deficiency (Nissen and Benson 1982), and that there existed genotypic variations among rice cultivars in the arsenic methylation ability (Zavala et al. 2008). Wu et al. (2002) attempted to measure the *in vitro* arsenic methylation activity in the cell extracts of bentgrass (*Agrostis capillaris*, formerly known as *Agrostis tenuis*) using *S*-[^{3}H-methyl]adenosyl-L-methionine (^{3}H-SAM) with either arsenite or arsenate as substrate. Their assay was based on the determination of the ^{3}H radioactivity; yet no direct evidence was provided demonstrating that ^{3}H was labelled onto the methylated arsenic compounds after the assay.

Recent studies have cast doubt on the ability of terrestrial plants to methylate As. Arao et al. (2011) showed that rice grown in non-sterile nutrient solution and supplied with arsenite contained considerable amounts of methylated arsenic (mainly DMA); DMA was also detected in the nutrient solution. However, addition of the antibacterial agent chloramphenicol to the nutrient solution markedly suppressed the DMA concentrations in both the nutrient solution and the rice plants, suggesting that methylated arsenic may have been synthesized by the bacteria in the growth medium. When rice plants were grown in axenic agar medium containing inorganic arsenic under either nutrient sufficient or deficient conditions, no methylated arsenic species were detected in the plants, whereas rice plants grown in soils under the same environmental conditions contained significant amounts of DMA (Lomax et al. 2012). Similarly, axenically grown plants of tomato and red clover, supplied with inorganic arsenic, had no detectable methylated arsenic. When MMA was supplied to rice plants in axenic culture, it was taken up by roots and partly reduced to MMA(III); however, there was no further methylation to DMA (Lomax et al. 2012). These results indicate that the plant species tested lack the ability to methylate arsenic, and that methylated arsenic species present in plants originate from the unsterile growth medium, where some microbes are able to convert inorganic arsenic to methylated species. It follows that genotypic variation in rice grain arsenic speciation (Liu et al. 2006; Norton et al. 2009; Pillai et al. 2010) reflects the variation in the root uptake or within plant transport of methylated arsenic, and/or the rhizosphere conditions that may influence microbial methylation. Also, the dominance of methylated arsenic in the rice grain produced in the southern states of the USA (Williams et al. 2007a; Zavala et al. 2008) is most likely a consequence of the specific soil conditions that promote microbial methylation, or the history of past use of methylated arsenicals rather than the difference in the *in planta* methylation ability.

Arsenic methylation in microbes involves both reduction and methylation steps (Bentley and Chasteen 2002). Plants are able to reduce arsenate (see Sect. 6.3.1) and pentavalent MMA(V) (Li et al. 2009a; Raab et al. 2005); however, methylation requires an arsenic methyltransferase, which may be absent in plants. A recent study has succeeded in transferring the arsenic methyltransferase gene *arsM* from the soil bacterium *Rhodopseudomonas palustris* to rice (Meng et al. 2011). The transgenic rice was able to methylate arsenic to MMA and DMA and to volatilize more arsenic than the untransformed plants, although the efficiency of arsenic methylation in the transgenic plants was still very low, with methylated arsenic species accounting for less than 1% of the total arsenic taken up by the plants.

6.4 Arsenic Sequestration in Plants

6.4.1 Iron Plaque

To cope with the anaerobic conditions in submerged soil, rice roots release O_2 to the rhizosphere through aerenchyma. Ferrous iron (Fe^{2+}), which is mobilized in the bulk soil solution, is oxidized to ferric iron (Fe^{2+}) forming precipitation of iron oxides/hydroxides on the root surface, the so-called iron plaque which has the characteristic orange colour (Fig. 6.9a). Analyses of the iron plaque on the

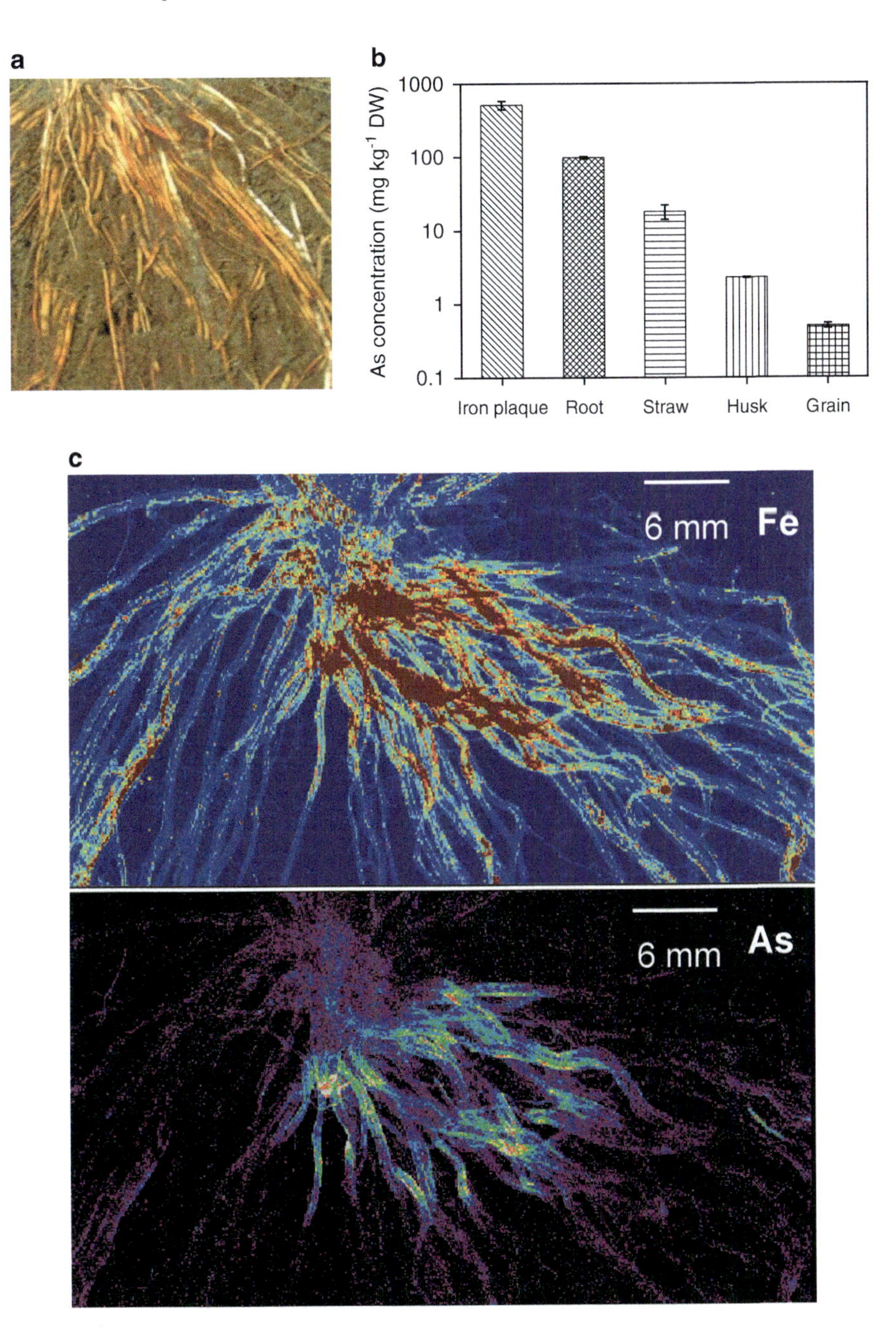

Fig. 6.9 Rice roots with iron plaque (**a**); decreasing gradient of arsenic concentration from iron plaque to rice grain (**b**); and synchrotron X-ray fluorescence images of iron and arsenic in a mature rice root system (**c**) (*Image* in (**a**) courtesy of Dr Wenju Liu. *Graph* in (**b**) was redrawn with permission from Liu et al. (2006). *Images* in (**c**) were reprinted with permission from Seyfferth et al. (2010). Copyright (2006, 2010) American Chemical Society)

roots of rice and another aquatic plant species (e.g. *Phalaris arundinacea*) show that the iron plaque consists mainly of ferrihydrite (60–100%) with goethite and siderite as minor components (Frommer et al. 2011; Hansel et al. 2001; Liu et al. 2006). As discussed in Chap. 5, iron oxides/hydroxides are strong sorbents for arsenic. The iron plaque on the root surface of rice therefore provides a strong sink for As. This is clearly shown in Fig. 6.9b; at rice maturity, iron plaque was found to contain a five times higher arsenic concentration than roots (the arsenic concentrations in both are based on the root dry weight) (Liu et al. 2006). It is also clear that there is a steep gradient in arsenic concentration from roots to straw, husk and grain. In general, arsenic is strongly associated with iron as revealed by X-ray fluorescence mapping of a mature rice root system (Fig. 6.9c), although there are also segments of roots where the two elements are not colocalized (Seyfferth et al. 2010). Chemical speciation analysis by X-ray absorption near-edge structure (XANES) show that the form of arsenic adsorbed by iron plaque is predominantly arsenate (Frommer et al. 2011; Liu et al. 2006; Seyfferth et al. 2010).

At the cellular level, arsenic, as well as phosphate and silicon, is strongly associated with iron plaque, which forms in the apoplast of the root epidermal cells and does not penetrate beyond the exodermal cells because of the Casparian strip (Fig. 6.10) (Moore et al. 2011).

Whilst iron plaque undoubtedly adsorbs a lot of arsenic, its effect on arsenic uptake and accumulation in shoots is less clear. On one hand, iron plaque may be a barrier preventing the entry of arsenic into the root cells. On the other hand, iron plaque serves as a strong sink for arsenic in the soil solution, and the strongly localized arsenic accumulation may become available to plant uptake; this may occur when the adsorbed arsenate on the iron plaque is replaced by phosphate. In short-term hydroponic experiments, Deng et al. (2010) reported decreases in the shoot arsenic concentration in plants with iron plaque compared with those without. In contrast, Liu et al. (2005) found little differences in the shoot arsenic concentration when arsenate was supplied, and even higher shoot arsenic concentration in the plants with iron plaque when arsenite was added. In short-term (30 min) uptake kinetic experiments, arsenate influx into rice roots appeared to be inhibited by the presence of iron plaque, whereas arsenite influx was enhanced (Chen et al. 2005). The formation of iron plaque on the rice root surface is not uniform; young roots and root tips usually have little iron plaque, and these roots or root segments are likely to be particularly important for the uptake of nutrients and contaminants (Seyfferth et al. 2010). There are also differences between primary thick roots and lateral fine roots; the former produced a concentric Fe and arsenic accumulation zone that extended from the root surface into the rhizosphere (~1 mm), whereas fine lateral roots produced a strong Fe and arsenic enrichment at the root surface with a particularly high As/Fe ratio (Frommer et al. 2011).

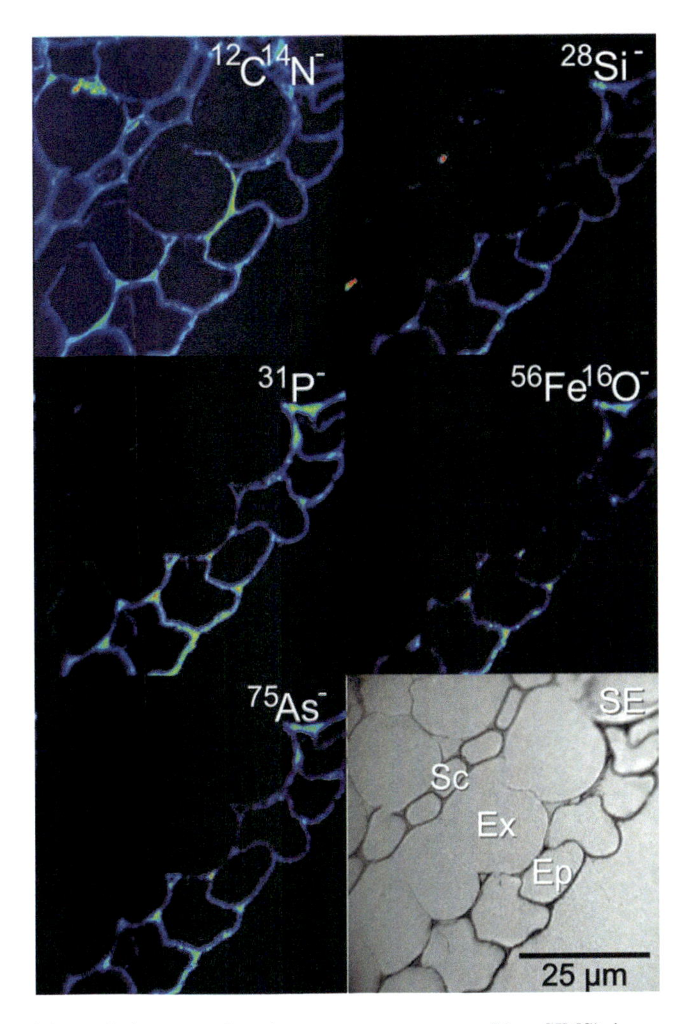

Fig. 6.10 High resolution secondary ion mass spectrometry (NanoSIMS) images of $^{12}C^{14}N^-$ (protein), $^{28}Si^-$, $^{31}P^-$, $^{56}Fe^{16}O^-$ and $^{75}As^-$ signals. Fe is detected as $^{56}Fe^{16}O^-$ with a lower sensitivity than for other elements. The *bottom right* image is a secondary electron (*SE*) image of the rice root cross section analysed by NanoSIMS. *Ep* epidermis, *Ex* exodermis, *Sc* sclerenchyma (Reprinted with permission from Moore et al. (2011). Copyright American Society of Plant Biologists)

6.4.2 Sequestration of Arsenic in the Vacuoles

It is thought that arsenic is sequestered in the vacuoles in the form of arsenite-thiol complexes (e.g. As(III)-PCs) (Zhao et al. 2009). The strong sequestration in the root cells explains the generally low mobility of arsenic from roots to shoots; As-hyperaccumulating ferns are exception because they form little As(III)-PCs

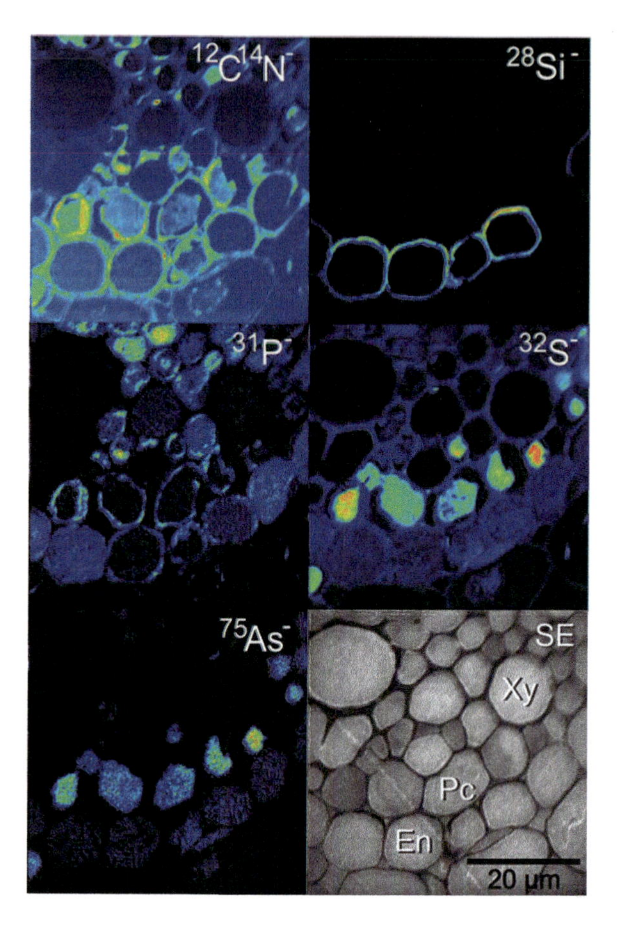

Fig. 6.11 NanoSIMS image of the stele region of a rice root (mature zone) showing arsenic accumulation in the vacuoles of pericycle and endodermis cells and a strong co-localisation with sulphur. Silicon is localized to the cell walls of the endodermis cells. *SE* secondary electron image, *En* endodermis, *Xy* xylem, *Pc* pericycle (Reprinted with permission from Moore et al. (2011). Copyright American Society of Plant Biologists)

(Raab et al. 2004a; Zhao et al. 2003). The transporters responsible for taking As(III)-PCs complexes across the tonoplast into the vacuoles have recently been identified in *Arabidopsis thaliana* (Song et al. 2010). These are the ATP binding cassette transporters AtABCC1 and AtABCC2. Mutant plants with both transporters non-functional are hypersensitive to arsenic, whereas overexpression of both in *Arabidopsis* increases arsenic tolerance.

A recent NanoSIMS study provided strong evidence of arsenic sequestration in the vacuoles of rice roots, especially pericycle cells and to a less extent endodermal cells (Fig. 6.11) (Moore et al. 2011). Moreover, there is a strong association of sulphur and arsenic in the vacuoles, consistent with the storage of arsenite-thiol complexes. Silicon has a very different cellular and subcellular distribution pattern, despite the same transport pathway for silicic acid and arsenite (see Sect. 6.2.2). This difference is due to the lack of tonoplast transporters for Si and, as a result, Si is highly efficiently loaded into the xylem for long-distance transport to the above-ground tissues of rice. Some accumulation of Si in the cell wall of the endodermal cells can be seen in Fig. 6.11.

6.5 Long-Distance Transport and Unloading of Arsenic in Rice Grain

6.5.1 *Xylem Transport*

With the exceptions of arsenic hyperaccumulator species, inorganic arsenic has a low mobility in the transport from roots to shoots. This is reflected by the low ratios of shoot to root arsenic concentrations (typically 0.1–0.3 in rice and <0.1 in other plants species) (Zhao et al. 2009). This is the case regardless whether roots are supplied with arsenate or arsenite. Despite arsenite and silicic acid sharing the same uptake pathway, silicic acid is much more mobile than arsenite in terms of the root-to-shoot translocation. In rice, a typical concentration ratio of Si in the xylem sap to that in the external medium is around 20 (Mitani and Ma 2005), whereas for arsenic this is only 0.3–0.6 (Zhao et al. 2009). Other plant species (except hyperaccumulators) have even lower arsenic mobility than rice with the xylem sap/external medium arsenic concentration ratio typically <0.1. In a hydroponic study with radioactive ^{73}As labelling, rice roots retained approximately 90% of the arsenite taken up, with only 10% ^{73}As being distributed to the shoots (Zhao et al. 2012). An important reason for the low mobility of arsenic is the complexation of arsenite with PCs and the sequestration of the complexes in the root vacuoles (see Sects. 6.3.2 and 6.4.2). The fact that rice has a larger arsenic mobility than other cereals (Su et al. 2010) can be attributed to the polar distribution of the silicic acid/arsenite influx and efflux transporters in the root exodermis and endodermis cells (Fig. 6.3), which ensures efficient delivery of the two molecules to the stele for xylem loading.

In the xylem sap collected from hydroponically grown rice, arsenite was the dominant arsenic species (typically 80–100%) even when plants were supplied with arsenate (Ma et al. 2008; Su et al. 2010; Zhao et al. 2009). These results were obtained when plants were exposed to arsenite or arsenate for 1 day. During the initial period of arsenate exposure (2 h), arsenate accounted for a larger proportion (55%) of the total arsenic in the xylem sap, suggesting that some arsenate was loaded into the xylem before being reduced to arsenite in the roots; however, this percentage decreased rapidly with the exposure time while the percentage of arsenite increased (Wu et al. 2011). In transgenic rice over-expressing the phosphate transporter OsPht1;8, not only the concentration of arsenate in the xylem sap increased dramatically, but also its proportion in the total arsenic increased (Wu et al. 2011). The constitutive over-expression of this transporter increases both the uptake and the xylem loading of arsenate. It is thus clear that both arsenate and arsenite can be loaded into the xylem, their relative proportions depending on the balance between arsenate influx, the reduction rate and the rate of loading into the xylem. In rice plants grown in flooded soil, arsenic speciation in the xylem sap was dominated by arsenite and DMA with very little arsenate. In flooded soil, the main forms of arsenic taken up and transported in rice are likely to be arsenite and DMA, with the phosphate/arsenate transport pathway contributing little to arsenic accumulation in rice (Wu et al. 2011).

Although the acidic pH (typically 5.5) of xylem sap favours the stability of As(III)-thiol complexes (Raab et al. 2004b), no such complexes were detected in the xylem sap of sunflower (Raab et al. 2005) or castor bean (*Ricinus communis*) (Ye et al. 2010). This is likely to be true in rice as well. In fact, *Arabidopsis thaliana* mutants lacking PCs had increased mobility of arsenic, consistent with the model that As(III) complexation with thiol compounds facilitates its sequestration in the root cells (Liu et al. 2010).

Despite slower uptake by roots, MMA and DMA are more mobile during root-to-shoot translocation than inorganic arsenic (Li et al. 2009a; Marin et al. 1992; Raab et al. 2007b; Ye et al. 2010). Xylem mobility increases in the order of inorganic As < MMA < DMA; the reason for this is unknown but may be related to the increasing hydrophobicity. Although MMA(V) was partly reduced to MMA(III) in the roots of rice and castor bean, only MMA(V) was detected in the xylem sap (Li et al. 2009a; Ye et al. 2010), possibly because the trivalent MMA(III) is complexed by thiols and sequestered in the root cells (Raab et al. 2005). In the xylem sap of both plant species exposed to DMA(V), only this arsenic species was detected (Li et al. 2009a; Ye et al. 2010).

6.5.2 Phloem Transport

The phloem transports sugars and amino acids from the source tissues (leaves) to the sink tissues (grain, roots and young leaves). Very little is known about the mechanisms of arsenic loading and unloading during phloem transport. In a study with castor bean, arsenic concentrations in the phloem sap were broadly comparable to those in the xylem sap from plants exposed to inorganic As, with the ratio varying from 0.7 to 3.9 (Ye et al. 2010). In both arsenite- and arsenate-treated plants, As(III) dominated in the phloem sap (70–94%) with the remainder being As(V). Moreover, there was no complexation of As(III) with thiol compounds, even though these were present at considerable concentrations in the phloem sap. An explanation for this lack of complexation is that As(III)-thiol complexes are unstable at the high pH (7.5–8.0) in the phloem sap. There has been no report of arsenic speciation in the phloem sap of rice.

Methylated arsenic species [MMA(V) and DMA(V)] are mobile during phloem transport. MMA(V) and DMA(V) were detected in the phloem sap of castor bean exposed to these arsenic species, and their concentrations were approximately seven- and three-fold larger than the corresponding concentrations in the xylem sap (Ye et al. 2010).

6.5.3 Transport of Arsenic into Rice Grain

Nutrients (including sugars, amino acids, minerals) are transported to the developing rice grain via the conducting tissue consisting of vascular bundles with xylem vessels and phloem. An important conducting tissue is the ovular vascular

Fig. 6.12 *Aniline blue* staining of the ovular vascular trace (OVT) in a young rice caryopsis (Reprinted with permission from Zheng et al. (2011))

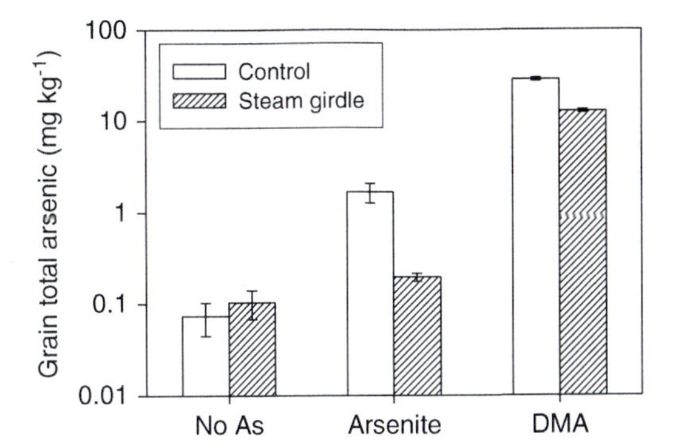

Fig. 6.13 Effect of steam girdling on the arsenic concentration in rice grain after 2-day feeding of arsenite (133 μM) or DMA (13.3 μM) through the cut stem below the flag leaf (Redrawn with permission from Carey et al. (2010). Copyright American Society of Plant Biologists)

trace (OVT) (Krishnan and Dayanandan 2003), shown in blue colour in Fig. 6.12 by aniline staining (Zheng et al. 2011). During the early stage of grain development, nutrients are transported from the OVT to the chalaza (also called pigment strand), into the nucellar projection and then into the aleurone cells and the endosperm. At later stages transport occurs from the OVT to the nucellar projection to the nucellar epidermis, from which they move centripetally towards the endosperm (Krishnan and Dayanandan 2003). There is a discontinuity in the symplastic transport route between the maternal tissues (OVT, chalaza, nucellar projection, nucellar epidermis) and the filial tissues (endosperm, aleurone layer and embryo). Nutrients exit the nucellar projection or nucellar epidermis into the apoplast before being taken up into the aleurone cells and the rest of the endosperm. This symplastic discontinuity may present a transport bottleneck for some of the minerals including inorganic arsenic.

Carey et al. (2010) evaluated the contribution of xylem versus phloem transport of inorganic arsenic and DMA to the developing rice caryopsis. They fed rice panicles, cut below the flag leaf, with arsenite or DMA together with the phloem transport marker rubidium and the xylem transport marker strontium; a group of panicles were steam girdled to destroy the phloem. Despite the concentration of DMA (13.3 μM) in the feeding solution being only one-tenth of that of arsenite (133 μM), rice grain accumulated 17-fold higher arsenic from the DMA treatment after 2-day feeding, indicating that DMA is much more mobile than arsenite (Fig. 6.13). Steam girdling decreased the grain arsenic concentration by 55% and 90% in the DMA and

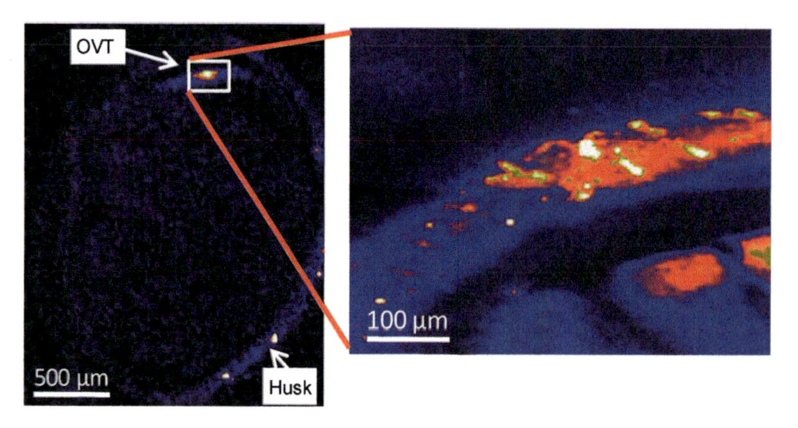

Fig. 6.14 Synchrotron X-Ray fluorescence mapping of arsenic in a cross section of a mature rice grain with husk. *OVT* ovular vascular trace (Reprinted with permission from Lombi et al. (2009))

arsenite treatments, respectively, suggesting that arsenite is delivered into rice grain mainly through phloem transport, whereas for DMA both the phloem and xylem transports make approximately equal contributions. Similarly, Zhao et al. (2012) showed that steam girdling decreased the translocation of ^{73}As-labelled arsenite to rice grain by 97%.

The difference in the mobility between inorganic and methylated arsenic species is also reflected by their different distribution patterns in rice grain. After a short-term (2 days) feeding through the cut stem, arsenite accumulated in the OVT region of rice caryopsis whereas DMA permeated into the external parts of the grain with some migration into the endosperm (Carey et al. 2010). In mature grain samples containing mainly inorganic arsenic, accumulation of arsenic in the OVT region is also apparent (Fig. 6.14) (Lombi et al. 2009; Meharg et al. 2008), whereas in a sample containing mainly (70%) DMA, arsenic was found to be concentrated in the sub-aleurone endosperm cells in association with the protein matrix (Moore et al. 2010). When brown rice is milled, the bran fraction contains a higher level of total arsenic than the polished rice (endosperm) (Sun et al. 2008). Moreover, bran arsenic is dominated by inorganic arsenic, whereas the endosperm has a larger percentage of DMA.

When arsenite, arsenate or DMA was fed to the flag leaves, DMA was highly efficiently translocated to the rice grain but arsenite and arsenate were not (Carey et al. 2011). In the study of Zhao et al. (2012), 2–3% of the ^{73}As labelled arsenite fed to the flag leaf was translocated to rice grain, suggesting that a small proportion of arsenite in leaves may be remobilized and transported to grain.

The high mobility of DMA in the xylem and phloem, contrasting with low mobility of inorganic arsenic, explains why the proportion of DMA in rice grain is much higher than in the vegetative tissues (Norton et al. 2010; Zheng et al. 2011). Although DMA represents a relatively small proportion of the total arsenic content in the whole plant of rice, it is preferentially distributed to the endosperm of rice grain.

6.6 Arsenic Toxicity and "Straighthead Disease"

6.6.1 Toxicity of Inorganic Arsenic

Both arsenate and arsenite are toxic to cellular metabolism. Arsenate, being a chemical analogue of phosphate, can replace phosphate in many biochemical reactions (Hughes 2002; Meharg and Hartley-Whitaker 2002). Arsenate uncouples *in vitro* formation of adenosine-5′-triphosphate (ATP) by a mechanism termed arsenolysis (Hughes 2002). Arsenate can react with adenosine-5′-diphosphate (ADP) to form ADP-arsenate, but this molecule is unstable as it hydrolyses easily due to the longer bond length of As-O than P-O (Hughes 2002). The mode of arsenite toxicity differs from that of arsenate. Due to its high affinity with thiol (–SH) groups, arsenite reacts readily with thiol-containing compounds, especially those with vicinal thiol groups, such as the dithiol lipoic acid. Lipoic acid is a cofactor of the enzyme pyruvate dehydrogenase, which oxidizes pyruvate to acetyl–CoA, a precursor to intermediates of the tricarboxylic acid (TCA) cycle. This enzyme is a sensitive target of arsenite toxicity (Chen et al. 2010; Hughes 2002).

Arsenic also causes toxicity by generating oxidative stresses. Reactive oxygen species may be generated during the redox reactions between arsenate and arsenite, or because of the depletion of reduced glutathione in response to arsenic exposure (Ahsan et al. 2008; Hartley-Whitaker et al. 2001; Requejo and Tena 2005).

Long-term irrigation of arsenic contaminated groundwater in Bangladesh has caused arsenic toxicity in rice (Khan et al. 2010; Panaullah et al. 2009). Panaullah et al. (2009) investigated the effect of arsenic accumulation in soil on rice production in a paddy field transect. They showed that as soil arsenic concentration increased from 10 to around 70 mg kg^{-1} near the inlet of irrigation channel, the grain yield of rice decreased linearly by up to 66% (Fig. 6.15). The average yield loss

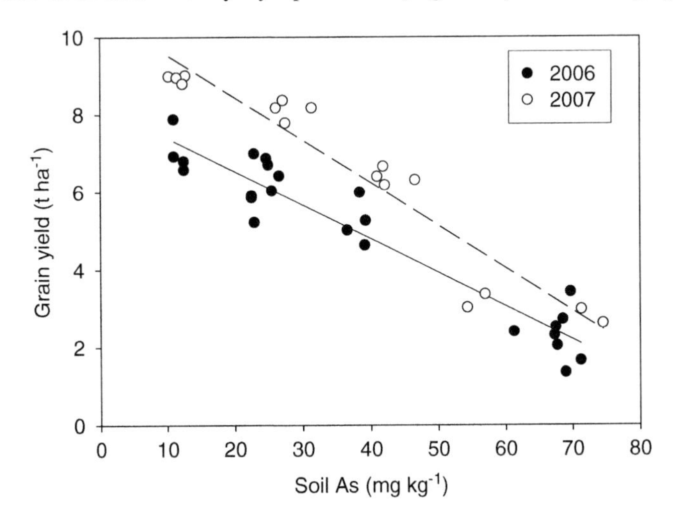

Fig. 6.15 Impact of soil arsenic accumulation on rice grain yield in 2 years (Redrawn with permission from Panaullah et al. (2009). Copyright Springer Science+Business Media)

over the 8 ha command area of the tubewell was estimated to be 16% (Panaullah et al. 2009). In a pot experiment, the growth of rice was inhibited when soil contained >15 mg As kg^{-1}, and severe toxicity symptoms were apparent when soil contained 60 mg As kg^{-1} (Khan et al. 2010). Growth inhibition was associated with straw arsenic concentrations of >5–10 mg kg^{-1} (Khan et al. 2010; Panaullah et al. 2009).

The relatively low level of arsenic in soil that can give rise to arsenic toxicity and yield losses in paddy rice is due to the mobilization, and thus enhanced bioavailability, of this toxic element under anaerobic conditions (see Chap. 5). Whilst irrigation of groundwater results in varying degrees of arsenic accumulation in paddy soil dependent on the extent of attenuation (see Chap. 5), accumulation of arsenic in soil to a toxic level can occur in some situations (Dittmar et al. 2010; Lu et al. 2009; Panaullah et al. 2009), which presents a serious challenge to agricultural sustainability in the region.

6.6.2 "Straighthead Disease"

Straighthead is a disease of rice that results in panicles remaining upright (hence straighthead) due to sterility and lack of grain fill (Belefant-Miller and Beaty 2007). The plants have little other side effects and, indeed, often have more lush, healthy foliage than normal grain yielding rice (Belefant-Miller and Beaty 2007). The sterility results in unfilled kernels and empty hulls distorted to a parrot-beak or crescent. This sterility differs considerably from sterility induced by alkali damage, diseases or drought (Belefant-Miller and Beaty 2007).

Straighthead can be caused by arsenic species, and arsenic induced straighthead has classically been found when rice is planted following a crop of cotton. Arsenical pesticides were commonly used as a boll weevil pesticide or defoliant in the South-Central US rice growing region (Belefant-Miller and Beaty 2007; Epps and Sturgis 1939; Williams et al. 2007a). However, straighthead is also found in Japan, South America and elsewhere (Belefant-Miller and Beaty 2007; Rasamivelona et al. 1995), and it is not just arsenic formulations that induce straighthead, but other triggers are more poorly understood (Belefant-Miller and Beaty 2007). Arsenic species come to the fore in straighthead as they can readily induce this disease, and thus arsenic has been studied in this respect (Agrama and Yan 2009; Gilmour and Wells 1980; Rahman et al. 2008; Rasamivelona et al. 1995; Wells and Gilmour 1977). This section will consider the relationship between arsenic species and straighthead by first reviewing the current state of knowledge. Potential hypotheses are then suggested to take forward our *enpasse* in understanding of arsenic induced straighthead, and straighthead in general.

6.6.2.1 Occurrence

Straighthead is most widely reported in the USA (Agrama and Yan 2009; Belefant-Miller and Beaty 2007; Gilmour and Wells 1980; Rahman et al. 2008; Rasamivelona et al. 1995; Wells and Gilmour 1977) and Japan (Iwamoto 1969; Odanaka et al. 1985).

It is notable that in areas of SE Asia where rice is irrigated with water containing high levels of inorganic arsenic, although yield reduction does occur due to arsenic toxicity (Panaullah et al. 2009), straighthead *per se* is not reported extensively, but there is a published report from Thailand (Rasamivelona et al. 1995). The prevalence of straighthead in the US, in particular, suggests that there may be environmental factors, i.e. edaphic/climatic, field management or cultivar differences coming into play. Rice varieties used in the US are highly susceptible to straighthead (Agrama and Yan 2009), resulting in breeding programmes to find resistant cultivars (Rasamivelona et al. 1995).

"Natural" (i.e. without specific chemical treatment) straighthead occurs in an unpredictable manor. Farmers in straighthead affected zones often drain and dry their paddy fields to reduce straighthead (Belefant-Miller and Beaty 2007), suggesting that straighthead is related to soil anaeorobism. Iwamoto (1969) surveying Japanese straighthead problems concluded that straighthead commonly occurred when paddy rice was cultured following sustained non-irrigated crop production, or when virgin soil was converted to rice cultivation. When paddy culture is sustained for 3–4 years patches of the field subject to straighthead gradually diminish. Straighthead was found not to be induced by climate or pest damage. Persistent flooding, low soil pH and high soil organic matter content have also been implicated in straighthead (Rahman et al. 2008).

Straighthead is not commonly reported in SE Asia. Rahman et al. (2008) investigated the potential of straighthead to be induced by inorganic arsenic, widely applied to Bangladeshi paddies as a result of agricultural groundwater contamination, using Bangladeshi soil and rice cultivars in arsenic amendment experiments. They found severe straighthead at soil application rates of 60 mg kg^{-1} arsenic. This is at the upper range of soil levels of arsenic in Bangladeshi soils, even in arsenic affected irrigation regions (Lu et al. 2009; Meharg and Rahman 2003; Williams et al. 2011). This straighthead potential for Bangladesh is worrying, and yield reduction has been observed for Bangladesh paddies under standard agronomic conditions (i.e. without experimental manipulation) with high arsenic irrigation waters and, therefore, high soil arsenic (Fig. 6.15), although this yield reduction was not explicitly stated in a straighthead context but a "reduction in the number of productive tillers, i.e. grain filled panicles" (Panaullah et al. 2009). Elsewhere, inorganic arsenic applied to cotton fields to treat boll weevil was found to induce straighthead (Reed and Sturgis 1936).

6.6.2.2 Physiology

Following an outbreak of straighthead that occurred in Stuttgart, Arkansas in a nutrient (N) experiment, inorganic elemental composition was analysed to see if there was any relationship between this composition and straighthead occurrence in three cultivars (Belefant-Miller and Beaty 2007). The outbreak was spontaneous, that is not induced by chemical treatment of the field. It was found that total arsenic concentration of plant parts (root, shoot, grain), or of soil, was related to straighthead. There was no overwhelming evidence that other elements were involved in straighthead response with respect to their plant concentrations. Nitrogen fertilization ameliorated straighthead to some degree, suggesting a role for nitrogen metabolism and/or nutrition. While straighthead affected the mineral composition of plant parts across a range of elements, it was only for magnesium and sodium that a consistent pattern was found, with both these elements being depleted in straighthead affected shoots. However, as Belafant-Miller and Beaty (2007) only presented ratio data rather than total elemental composition, further interpretation with respect to nutrient status was not possible.

Phosphorus, potassium, iron and sulphate have also been implicated in "natural" straighthead (Iwamoto 1969), while a wide range of soil amendments can induce straighthead including permanganate, boron, sulphates, iron, copper, zinc, aluminium (Belefant-Miller and Beaty 2007), as well as arsenic salts. The fact that straighthead is readily induced by a range of soil amendments suggests that these compounds are profoundly altering key nutrient(s) availability or cycling within reduced paddy soil environments, leading to straighthead. It is also likely that the impact of these amendments will be variable among different soils, thus making the underlying causes of straighthead difficult to pin down. This was the conclusion of Iwamoto (1969), who speculated that it was the response of plant thiol metabolism to local soil conditions that led to straighthead.

The role of plant growth regulators, ethylene and indoleacetic acid (IAA), in MMA induced straighthead was investigated, and it was concluded that they had no role in the disease (Horton et al. 1983). Under experimental manipulation nitrogen was found to ameliorate straighthead induced by MMA, but the effect was only slight (Yan et al. 2005). The role of nitrogen nutrition on ameliorating straighthead had also been observed under field conditions (Belefant-Miller and Beaty 2007).

Field experiments showed that straighthead could persist for 2 years after the application of MMA, but the impact of the disease could be decreased by a midseason draining and drying water management regime, again implicating conditions induced by anaerobism as a causal factor (Gilmour and Wells 1980).

6.6.2.3 Genetics

Straighthead resistance and sensitivity is widely found, particularly in US germplasm (Agrama and Yan 2009; Rasamivelona et al. 1995; Yan et al. 2005) where straighthead disease is an economic problem, but has also has been found in ~15% of screened Chinese cultivars (Yan et al. 2005). The heritability of straighthead resistance is high, with little genotype by environment interactions (Rasamivelona et al. 1995). Association mapping was performed on 547 cultivars from the United States Department of Agriculture (USDA) core rice collection treated with MMA and then scored for straighthead impact (Agrama and Yan 2009). Out of 75 markers only 2 were associated with straighthead resistance, whereas 6 markers, accounting for 35% of phenotypic variation, were associated with the occurrence of straighthead.

It is relatively straightforward to select and breed straighthead resistance (Agrama and Yan 2009; Rasamivelona et al. 1995; Yan et al. 2005), and along with field management of midseason draining, perhaps this has negated the need to resolve the physiological basis of straighthead resistances. Such toxicant resistances in themselves have inherent interest as model evolutionary systems (Macnair 1993), and it is interesting that MMA has such a strong effect just on the plants reproductive capabilities, which contrasts with general arsenate/arsenite resistances observed in plants (Meharg and Hartley-Whitaker 2002) and indeed for rice itself (Lou-Hing et al. 2011). More practically however, adaptations to MMA, and other straighthead inducing scenarios, may prove beneficial to increasing rice yield, particularly for regions where straighthead may be an emerging threat (Rahman et al. 2008). Fundamental underlying knowledge of the genetic basis of straighthead can only improve optimization of breeding and of field management to decrease this disease. Agrama and Yan (2009) used an association panel to start to identify resistance and sensitivity loci. With ever more powerful association mapping populations, in terms of density of markers (Huang et al. 2010) and, therefore, the specificity of identifying loci location, and with such a strong phenotype (sterility induced by MMA), identifying the genes involved is highly tractable.

6.6.2.4 Re-evaluation of the Role of Arsenic in Straighthead

The understanding of arsenic biogeochemical cycling in paddy soils and plant assimilation and metabolism has increased exponentially over recent years due to health concerns regarding inorganic arsenic in rice grain (see Chaps. 5 and 6). Marin et al. (1992, 1993) stated that considerations regarding the toxicology of arsenic, including straighthead, had so far entirely focused on total arsenic in plant and soil. They also noted that very little was known about the mode of toxicity of MMA and DMA to rice and other plants, despite these compounds being widely used as agricultural pesticides. They are thought to impact photosynthesis and carbohydrate allocation (Marin et al. 1992, 1993). These methylated species appear not to be demethylated in plants, so the products of demethylation do not appear to be part of their toxicology. It has been noted for sometime (Marin et al. 1992, 1993) that methylated species are more effectively translocated to the shoot, as compared to inorganic species, and this efficient shoot translocation may affect their toxicology. Carey et al. (2011) observed direct effects of MMA and DMA on water transpiration in rice, where arsenic species were fed directly into flag leaf. In contrast, inorganic species did not affect transpiration. This effect on transpiration is not surprising given that MMA and DMA are widely used as cotton leaf desiccants (Marin et al. 1992, 1993). It is MMA that is often used to induce straighthead rather than DMA, and there is little regarding the relationship between DMA and straighthead in the literature, even though the herbicidal properties of these compounds are similar (Marin et al. 1992, 1993). DMA is found widely in soils naturally or after inorganic or MMA treatment (Marin et al. 1993; Mestrot et al. 2009, 2011; Takamatsu et al. 1982; Williams et al. 2011; Woolson et al. 1982) (see Chap. 5).

Speciation studies on rice have shown that DMA is a major component of grain total arsenic (see Chap. 2), but typically accounts for only ~5% of shoot arsenic (Norton et al. 2010). MMA in grain is typically lower than detection limits but can be found on occasions at trace concentrations (Williams et al. 2005). DMA and MMA are particularly efficiently exported to above ground tissues in rice (Li et al. 2009a; Marin et al. 1992, 1993; Raab et al. 2007b), as well as to grain (Carey et al. 2011; Norton et al. 2010). DMA present in grain varies greatly globally, with US rice having much higher presence of DMA compared to, say, India, Bangladesh and China (Meharg et al. 2009; Williams et al. 2005; Zavala et al. 2008). This DMA is soil derived (see Sect. 6.3.3) and regional differences in DMA in rice grain must reflect differences in soil edaphic/climatic factors, especially the microbial community. This is typified by a study by Norton et al. (2012) who grew ~200 rice cultivars on a Bangladeshi and Chinese site, showing that the Chinese site had much higher levels of DMA in grain than the Bangladeshi site, identifying that grain DMA was under environmental, rather than genetic, control. As DMA occurs "naturally" from microbial methylation of inorganic arsenic (see Chap. 5), it may be responsible for "natural" straighthead (Belefant-Miller and Beaty 2007). The link between DMA and straighthead remains a hypothesis and needs future experimental testing, but there is some evidence supporting this link. The most prevalent straighthead area, south-central USA, also produces rice grain with higher DMA concentrations than other regions of the world (Williams et al. 2007a; Zavala et al. 2008). In a pot experiment, an addition of arsenate or arsenite to soil increased spikelet sterility in rice grown under flooded conditions, but the grain arsenic was dominated by DMA (~70%), suggesting the possibility of DMA causing sterility (Xu et al. 2008). In hydroponic experiments, exposure of rice plants to DMA caused typical straighthead symptoms (e.g. beak-shaped husk and abnormal floral organs, Fig. 6.16) and substantially increased spikelet sterility, whereas exposure to inorganic arsenic did not (Liu WJ, Zhao FJ, unpublished; Zheng M, Zhu YG, personal communications, 2011). One difference between soil and hydroponic experiments is that inorganic arsenic can be converted to DMA by soil microorganisms, but this process is very limited in hydroponic culture. Also of note, Williams et al. (2005) conducted experiments on straighthead sensitive and insensitive US rice cultivars and found that on soils treated with inorganic arsenic that grain and shoot nitrogen, particularly the later, increased dramatically on arsenic treatment, suggesting that nitrogen nutrition is an important response in arsenic affected plants. Belefant-Miller and Beaty (2007) had found that nitrogen fertilization led to a slight alleviation of straighthead. It was notable that a high percentage (up to 56%) of grain arsenic was DMA in the experiments of Williams et al. (2005), again even though the plants were only treated with inorganic arsenic, for both control and arsenic treated plants.

This is not to say that MMA is not a major inducer of straighthead, especially historically where it was widely used in agricultural environments where rice was grown in the US, in particular. However, it appears that it is the initial methylation of inorganic arsenic that is the limitation in methylation, and once methylated species are present, further methylation is more rapid (Mestrot et al. 2011), i.e. naturally produced MMA is readily converted to DMA, and it is DMA that is mainly attained

Fig. 6.16 Normal floret and flower organs of rice in untreated control (**a**); and beak-shaped florets and abnormal flower organs in DMA treated plants (**b**–**h**). Note the increased numbers of ovaries (*arrow head*) and pistils (*red arrow*), and irregular 'lumps' (*starred*) on the surfaces of some ovaries (**e**–**g**). Abbreviations in (**a**): *le* lemma, *pa* palea, *st* stamen, *pi* pistil, *ov* ovary, *gl* glume, *lo* lodicule (Images were kindly provided by Dr Maozhong Zheng)

by the plant. This makes sense as there is no evidence of arsenic methylation in rice plant (see Sect. 6.3.3), and MMA is readily translocated, as effectively as DMA, to shoots and grain (Carey et al. 2011; Raab et al. 2007b).

It is now becoming clear that soil amendments such as organic matter, implicated in causing straighthead (Iwamoto 1969), can increase DMA content of paddy soil (Mestrot et al. 2009, 2011), while natural dissolved organic matter (DOM) has a similar effect (Williams et al. 2011). Furthermore, inorganic amendments such as phosphate and silicic acid can also lead to enhanced DMA concentration in grain (Li et al. 2009b; Wu et al. 2011). It is thought that the mechanism behind this is that the DOM or inorganic ligands (e.g. phosphate and silicic acid) lead to desorption of inorganic arsenic from the adsorption sites in the soil, releasing this into porewater where it may be methylated by soil microflora (Mestrot et al. 2011; Williams et al. 2011; Wu et al. 2011). Furthermore, anaerobic conditions may develop more rapidly in soils with a high concentration of easily decomposable organic matter, promoting the mobilization of inorganic arsenic and subsequent methylation. Thus, the fact that straighthead can be caused by local soil variation in "natural" conditions, or induced by a wide range of organic and inorganic amendments (Belefant-Miller and Beaty 2007; Iwamoto 1969), may again point to straighthead being caused by DMA.

Realizing that rice readily assimilates "natural" and chemically stimulated DMA production negates most historic work on the role of arsenic in straighthead as only total arsenic was considered, as typified by Belefant-Millar and Beaty (2007). It is now apparent that arsenic speciation in soils and plants could be central to understanding straighthead. This must be a priority in future research. In terms of genetic variation in straighthead sensitivity/resistance, it could be due to either different uptake/translocation efficiency of DMA or different capacity to detoxify DMA.

References

Abbas MHH, Meharg AA (2008) Arsenate, arsenite and dimethyl arsinic acid (DMA) uptake and tolerance in maize (*Zea mays* L.). Plant Soil 304:277–289

Abedin MJ, Feldmann J, Meharg AA (2002) Uptake kinetics of arsenic species in rice plants. Plant Physiol 128:1120–1128

Agrama HA, Yan WG (2009) Association mapping of straighthead disorder induced by arsenic in *Oryza sativa*. Plant Breed 128:551–558

Ahsan N, Lee DG, Alam I, Kim PJ, Lee JJ, Ahn YO, Kwak SS, Lee IJ, Bahk JD, Kang KY, Renaut J, KomatsU S, Lee BH (2008) Comparative proteomic study of arsenic-induced differentially expressed proteins in rice roots reveals glutathione plays a central role during As stress. Proteomics 8:3561–3576

Ai PH, Sun SB, Zhao JN, Fan XR, Xin WJ, Guo Q, Yu L, Shen QR, Wu P, Miller AJ, Xu GH (2009) Two rice phosphate transporters, OsPht1;2 and OsPht1;6, have different functions and kinetic properties in uptake and translocation. Plant J 57:798–809

Arao T, Kawasaki A, Baba K, Matsumoto S (2011) Effects of arsenic compound amendment on arsenic speciation in rice grain. Environ Sci Technol 45:1291–1297

Asher CJ, Reay PF (1979) Arsenic uptake by barley seedlings. Aust J Plant Physiol 6:459–466

Bansal A, Sankararamakrishnan R (2007) Homology modeling of major intrinsic proteins in rice, maize and Arabidopsis: comparative analysis of transmembrane helix association and aromatic/arginine selectivity filters. BMC Struct Biol 7:27

Barber SA (1984) Soil nutrient bioavailability, A mechanistic approach. Wiley, New York

Belefant-Miller H, Beaty T (2007) Distribution of arsenic and other minerals in rice plants affected by natural straighthead. Agron J 99:1675–1681

Bentley R, Chasteen TG (2002) Microbial methylation of metalloids: arsenic, antimony, and bismuth. Microbiol Mol Biol Rev 66:250–271

Bienert GP, Schuessler MD, Jahn TP (2008a) Metalloids: essential, beneficial or toxic? Major intrinsic proteins sort it out. Trends Biochem Sci 33:20–26

Bienert GP, Thorsen M, Schüssler MD, Nilsson HR, Wagner A, Tamás MJ, Jahn TP (2008b) A subgroup of plant aquaporins facilitate the bi-directional diffusion of $As(OH)_3$ and $Sb(OH)_3$ across membranes. BMC Biol 6:26

Bleeker PM, Hakvoort HWJ, Bliek M, Souer E, Schat H (2006) Enhanced arsenate reduction by a CDC25-like tyrosine phosphatase explains increased phytochelatin accumulation in arsenate-tolerant *Holcus lanatus*. Plant J 45:917–929

Bogdan K, Schenk MK (2008) Arsenic in Rice (*Oryza sativa* L.) related to dynamics of arsenic and silicic acid in daddy soils. Environ Sci Technol 42:7885–7890

Carbonell-Barrachina AA, Aarabi MA, DeLaune RD, Gambrell RP, Patrick WH (1998) The influence of arsenic chemical form and concentration on *Spartina patens* and *Spartina alterniflora* growth and tissue arsenic concentration. Plant Soil 198:33–43

Carbonell-Barrachina AA, Burlo F, Valero D, Lopez E, Martinez-Romero D, Martinez-Sanchez F (1999) Arsenic toxicity and accumulation in turnip as affected by arsenic chemical speciation. J Agric Food Chem 47:2288–2294

Carey AM, Scheckel KG, Lombi E, Newville M, Choi Y, Norton GJ, Charnock JM, Feldmann J, Price AH, Meharg AA (2010) Grain unloading of arsenic species in rice. Plant Physiol 152:309–319

Carey AM, Norton GJ, Deacon C, Scheckel KG, Lombi E, Punshon T, Guerinot ML, Lanzirotti A, Newville M, Choi Y, Price AH, Meharg AA (2011) Phloem transport of arsenic species from flag leaf to grain during grain filling. New Phytol 192:87–98

Chen Z, Zhu YG, Liu WJ, Meharg AA (2005) Direct evidence showing the effect of root surface iron plaque on arsenite and arsenate uptake into rice (*Oryza sativa*) roots. New Phytol 165:91–97

Chen W, Chi Y, Taylor NL, Lambers H, Finnegan PM (2010) Disruption of ptLPD1 or ptLPD2, genes that encode isoforms of the plastidial lipoamide dehydrogenase, confers arsenate hypersensitivity in *Arabidopsis*. Plant Physiol 153:1385–1397

Cobbett C, Goldsbrough P (2002) Phytochelatins and metallothioneins: roles in heavy metal detoxification and homeostasis. Annu Rev Plant Biol 53:159–182

Delnomdedieu M, Basti MM, Otvos JD, Thomas DJ (1994) Reduction and binding of arsenate and dimethylarsinate by glutathione – A magnetic resonance study. Chem Biol Interact 90:139–155

Deng D, Wu SC, Wu FY, Deng H, Wong MH (2010) Effects of root anatomy and Fe plaque on arsenic uptake by rice seedlings grown in solution culture. Environ Pollut 158:2589–2595

Dhankher OP, Li YJ, Rosen BP, Shi J, Salt D, Senecoff JF, Sashti NA, Meagher RB (2002) Engineering tolerance and hyperaccumulation of arsenic in plants by combining arsenate reductase and gamma-glutamylcysteine synthetase expression. Nat Biotechnol 20:1140–1145

Dhankher OP, Rosen BP, McKinney EC, Meagher RB (2006) Hyperaccumulation of arsenic in the shoots of *Arabidopsis* silenced for arsenate reductase (ACR2). Proc Natl Acad Sci USA 103:5413–5418

Dittmar J, Voegelin A, Roberts LC, Hug SJ, Saha GC, Ali MA, Badruzzaman ABM, Kretzschmar R (2010) Arsenic accumulation in a paddy field in Bangladesh: seasonal dynamics and trends over a three-year monitoring period. Environ Sci Technol 44:2925–2931

Duan GL, Zhou Y, Tong YP, Mukhopadhyay R, Rosen BP, Zhu YG (2007) A CDC25 homologue from rice functions as an arsenate reductase. New Phytol 174:311–321

Ellis DR, Gumaelius L, Indriolo E, Pickering IJ, Banks JA, Salt DE (2006) A novel arsenate reductase from the arsenic hyperaccumulating fern *Pteris vittata*. Plant Physiol 141: 1544–1554

Epps EA, Sturgis MB (1939) Arsenic compounds toxic to rice. Soil Sci Soc Am Proc 4:215–218

Forrest KL, Bhave M (2007) Major intrinsic proteins (MIPs) in plants: a complex gene family with major impacts on plant phenotype. Funct Integr Genomics 7:263–289

Frommer J, Voegelin A, Dittmar J, Marcus MA, Kretzschmar R (2011) Biogeochemical processes and arsenic enrichment around rice roots in paddy soil: results from micro-focused X-ray spectroscopy. Eur J Soil Sci 62:305–317

Geiszinger A, Goessler W, Kosmus W (2002) Organoarsenic compounds in plants and soil on top of an ore vein. Appl Organomet Chem 16:245–249
Gilmour JT, Wells BR (1980) Residual effects of MSMA on sterility in rice cultivars. Agron J 72:1066–1067
González E, Solano R, Rubio V, Leyva A, Paz-Ares J (2005) PHOSPHATE TRANSPORTER TRAFFIC FACILITATOR1 is a plant-specific SEC12-related protein that enables the endoplasmic reticulum exit of a high-affinity phosphate transporter in *Arabidopsis*. Plant Cell 17:3500–3512
Guo W, Zhu YG, Liu WJ, Liang YC, Geng CN, Wang SG (2007) Is the effect of silicon on rice uptake of arsenate (As-V) related to internal silicon concentrations, iron plaque and phosphate nutrition? Environ Pollut 148:251–257
Guo JB, Dai XJ, Xu WZ, Ma M (2008) Overexpressing *GSH1* and *AsPCS1* simultaneously increases the tolerance and accumulation of cadmium and arsenic in *Arabidopsis thaliana*. Chemosphere 72:1020–1026
Ha SB, Smith AP, Howden R, Dietrich WM, Bugg S, O'Connell MJ, Goldsbrough PB, Cobbett CS (1999) Phytochelatin synthase genes from *Arabidopsis* and the yeast *Schizosaccharomyces pombe*. Plant Cell 11:1153–1163
Hansel CM, Fendorf S, Sutton S, Newville M (2001) Characterization of Fe plaque and associated metals on the roots of mine-waste impacted aquatic plants. Environ Sci Technol 35:3863–3868
Hartley-Whitaker J, Ainsworth G, Meharg AA (2001) Copper- and arsenate-induced oxidative stress in *Holcus lanatus* L. clones with differential sensitivity. Plant Cell Environ 24:713–722
Horton DK, Frans RE, Cothren T (1983) MSMA-induced straighthead in rice (*Oryza sativa*) and effect upon metabolism and yield. Weed Sci 31:648–651
Howden R, Andersen CR, Goldsbrough PB, Cobbett CS (1995) A cadmium-sensitive, glutathione deficient mutant of *Arabidopsis thaliana*. Plant Physiol 107:1067–1073
Huang XH, Wei XH, Sang T, Zhao QA, Feng Q, Zhao Y, Li CY, Zhu CR, Lu TT, Zhang ZW, Li M, Fan DL, Guo YL, Wang A, Wang L, Deng LW, Li WJ, Lu YQ, Weng QJ, Liu KY, Huang T, Zhou TY, Jing YF, Li W, Lin Z, Buckler ES, Qian QA, Zhang QF, Li JY, Han B (2010) Genome-wide association studies of 14 agronomic traits in rice landraces. Nat Genet 42:961–967
Hughes MF (2002) Arsenic toxicity and potential mechanisms of action. Toxicol Lett 133:1–16
Isayenkov SV, Maathuis FJM (2008) The *Arabidopsis thaliana* aquaglyceroporin AtNIP7;1 is a pathway for arsenite uptake. FEBS Lett 582:1625–1628
Iwamoto R (1969) Straighthead of rice plants effected by functional abnormality of thiol-compound metabolism. Memo Tokyo Univ Agriculture 13:62–80
Jia HF, Ren HY, Gu M, Zhao JN, Sun SB, Zhang X, Chen JY, Wu P, Xu GH (2011) The phosphate transporter gene *OsPht1;8* is involved in phosphate homeostasis in rice. Plant Physiol 156:1164–1175
Kamiya T, Tanaka M, Mitani N, Ma JF, Maeshima M, Fujiwara T (2009) NIP1;1, an aquaporin homolog, determines the arsenite sensitivity of *Arabidopsis thaliana*. J Biol Chem 284:2114–2120
Khan MA, Stroud JL, Zhu YG, McGrath SP, Zhao FJ (2010) Arsenic bioavailability to rice is elevated in Bangladeshi paddy soils. Environ Sci Technol 44:8515–8521
Krishnan S, Dayanandan P (2003) Structural and histochemical studies on grain-filling in the caryopsis of rice (*Oryza sativa* L.). J Biosci 28:455–469
Kuehnelt D, Lintschinger J, Goessler W (2000) Arsenic compounds in terrestrial organisms. IV. Green plants and lichens from an old arsenic smelter site in Austria. Appl Organomet Chem 14:411–420
Lee RB (1982) Selectivity and kinetics of ion uptake by barley plants following nutrient deficiency. Ann Bot 50:429–449
Li YJ, Dhankher OP, Carreira L, Lee D, Chen A, Schroeder JI, Balish RS, Meagher RB (2004) Overexpression of phytochelatin synthase in *Arabidopsis* leads to enhanced arsenic tolerance and cadmium hypersensitivity. Plant Cell Physiol 45:1787–1797
Li RY, Ago Y, Liu WJ, Mitani N, Feldmann J, McGrath SP, Ma JF, Zhao FJ (2009a) The rice aquaporin Lsi1 mediates uptake of methylated arsenic species. Plant Physiol 150:2071–2080

Li RY, Stroud JL, Ma JF, McGrath SP, Zhao FJ (2009b) Mitigation of arsenic accumulation in rice with water management and silicon fertilization. Environ Sci Technol 43:3778–3783

Liu WJ, Zhu YG, Smith FA (2005) Effects of iron and manganese plaques on arsenic uptake by rice seedlings (*Oryza sativa* L.) grown in solution culture supplied with arsenate and arsenite. Plant Soil 277:127–138

Liu WJ, Zhu YG, Hu Y, Williams PN, Gault AG, Meharg AA, Charnock JM, Smith FA (2006) Arsenic sequestration in iron plaque, its accumulation and speciation in mature rice plants (*Oryza sativa* L.). Environ Sci Technol 40:5730–5736

Liu WJ, Wood BA, Raab A, McGrath SP, Zhao FJ, Feldmann J (2010) Complexation of arsenite with phytochelatins reduces arsenite efflux and translocation from roots to shoots in Arabidopsis. Plant Physiol 152:2211–2221

Lomax C, Liu WJ, Wu LY, Xue K, Xiong J, Zhou JZ, McGrath SP, Meharg AA, Miller AJ, Zhao FJ (2012) Methylated arsenic species in plants originate from soil microorganisms. New Phytol doi: 10.1111/j.1469-8137.2011.03956.x

Lombi E, Scheckel KG, Pallon J, Carey AM, Zhu YG, Meharg AA (2009) Speciation and distribution of arsenic and localization of nutrients in rice grains. New Phytol 184:193–201

Lou-Hing D, Zhang B, Price AH, Meharg AA (2011) Effects of phosphate on arsenate and arsenite sensitivity in two rice (*Oryza sativa* L.) cultivars of different sensitivity. Environ Exp Bot 72:47–52

Lu Y, Adomako EE, Solaiman ARM, Islam MR, Deacon C, Williams PN, Rahman G, Meharg AA (2009) Baseline soil variation is a major factor in arsenic accumulation in Bengal delta paddy rice. Environ Sci Technol 43:1724–1729

Ma JF, Yamaji N (2006) Silicon uptake and accumulation in higher plants. Trends Plant Sci 11:392–397

Ma JF, Tamai K, Yamaji N, Mitani N, Konishi S, Katsuhara M, Ishiguro M, Murata Y, Yano M (2006) A silicon transporter in rice. Nature 440:688–691

Ma JF, Yamaji N, Mitani N, Tamai K, Konishi S, Fujiwara T, Katsuhara M, Yano M (2007) An efflux transporter of silicon in rice. Nature 448:209–212

Ma JF, Yamaji N, Mitani N, Xu XY, Su YH, McGrath SP, Zhao FJ (2008) Transporters of arsenite in rice and their role in arsenic accumulation in rice grain. Proc Natl Acad Sci USA 105:9931–9935

Macnair MR (1993) The genetics of metal tolerance in vascular plants. New Phytol 124:541–559

Marin AR, Masscheleyn PH, Patrick WH (1992) The influence of chemical form and concentration of arsenic on rice growth and tissue arsenic concentration. Plant Soil 139:175–183

Marin AR, Pezeshki SR, Masschelen PH, Choi HS (1993) Effect of dimethylarsenic Acid (DMAA) on growth, tissue arsenic, and photosynthesis of rice plants. J Plant Nutr 16:865–880

Maurel C, Verdoucq L, Luu DT, Santoni V (2008) Plant aquaporins: membrane channels with multiple integrated functions. Ann Rev Plant Biol 59:595–624

Meharg AA, Hartley-Whitaker J (2002) Arsenic uptake and metabolism in arsenic resistant and nonresistant plant species. New Phytol 154:29–43

Meharg AA, Jardine L (2003) Arsenite transport into paddy rice (*Oryza sativa*) roots. New Phytol 157:39–44

Meharg AA, Macnair MR (1992) Suppression of the high-affinity phosphate uptake system: a mechanism of arsenate tolerance in *Holcus lanatus* L. J Exp Bot 43:519–524

Meharg AA, Rahman M (2003) Arsenic contamination of Bangladesh paddy field soils: Implications for rice contribution to arsenic consumption. Environ Sci Technol 37:229–234

Meharg AA, Lombi E, Williams PN, Scheckel KG, Feldmann J, Raab A, Zhu YG, Islam R (2008) Speciation and localization of arsenic in white and brown rice grains. Environ Sci Technol 42:1051–1057

Meharg AA, Williams PN, Adomako E, Lawgali YY, Deacon C, Villada A, Cambell RCJ, Sun G, Zhu YG, Feldmann J, Raab A, Zhao FJ, Islam R, Hossain S, Yanai J (2009) Geographical variation in total and inorganic arsenic content of polished (white) rice. Environ Sci Technol 43:1612–1617

Meng XY, Qin J, Wang LH, Duan GL, Sun GX, Wu HL, Chu CC, Ling HQ, Rosen BP, Zhu YG (2011) Arsenic biotransformation and volatilization in transgenic rice. New Phytol 191:49–56

Mestrot A, Uroic MK, Plantevin T, Islam MR, Krupp EM, Feldmann J, Meharg AA (2009) Quantitative and qualitative trapping of arsines deployed to assess loss of volatile arsenic from paddy soil. Environ Sci Technol 43:8270–8275

Mestrot A, Feldmann J, Krupp EM, Hossain MS, Roman-Ross G, Meharg AA (2011) Field fluxes and speciation of arsines emanating from soils. Environ Sci Technol 45:1798–1804

Mihucz VG, Tatar E, Virag I, Cseh E, Fodor F, Zaray G (2005) Arsenic speciation in xylem sap of cucumber (*Cucumis sativus* L.). Anal Bioanal Chem 383:461–466

Mitani N, Ma JF (2005) Uptake system of silicon in different plant species. J Exp Bot 56:1255–1261

Mitani N, Yamaji N, Ma JF (2008) Characterization of substrate specificity of a rice silicon transporter, Lsi1. Pflugers Arch 456:679–686

Mitani N, Chiba Y, Yamaji N, Ma JF (2009) Identification and characterization of maize and barley Lsi2-like silicon efflux transporters reveals a distinct silicon uptake system from that in rice. Plant Cell 21:2133–2142

Mitani-Ueno N, Yamaji N, Zhao FJ, Ma JF (2011) The aromatic/arginine selectivity filter of NIP aquaporins plays a critical role in substrate selectivity for silicon, boron, and arsenic. J Exp Bot 62:4391–4398

Moore KL, Schroder M, Lombi E, Zhao FJ, McGrath SP, Hawkesford MJ, Shewry PR, Grovenor CRM (2010) NanoSIMS analysis of arsenic and selenium in cereal grain. New Phytol 185:434–445

Moore KL, Schröder M, Wu ZC, Martin BGH, Hawes CR, McGrath SP, Hawkesford MJ, Ma JF, Zhao FJ, Grovenor CRM (2011) NanoSIMS analysis reveals contrasting patterns of arsenic and silicon localization in rice roots. Plant Physiol 156:913–924

Nissen P, Benson AA (1982) Arsenic metabolism in fresh-water and terrestrial plants. Physiol Plant 54:446–450

Norton GJ, Lou-Hing DE, Meharg AA, Price AH (2008) Rice-arsenate interactions in hydroponics: whole genome transcriptional analysis. J Exp Bot 59:2267–2276

Norton GJ, Duan GL, Dasgupta T, Islam MR, Lei M, Zhu YG, Deacon CM, Moran AC, Islam S, Zhao FJ, Stroud JL, McGrath SP, Feldmann J, Price AH, Meharg AA (2009) Environmental and genetic control of arsenic accumulation and speciation in rice grain: comparing a range of common cultivars grown in contaminated sites across Bangladesh, China, and India. Environ Sci Technol 43:8381–8386

Norton GJ, Pinson SRM, Alexander J, Mckay S, Hansen H, Duan GL, Islam MR, Islam S, Stroud JL, Zhao FJ, McGrath SP, Zhu YG, Lahner B, Yakubova E, Guerinot ML, Tarpley L, Eizenga GC, Salt DE, Meharg AA, Price AH (2012) Variation in grain arsenic assessed in a diverse panel of rice (Oryza sativa) grown in multiple sites. New Phytol doi: 10.1111/j.1469-8137.2011.03983.x

Norton GJ, Islam MR, Duan GL, Lei M, Zhu YG, Deacon CM, Moran AC, Islam S, Zhao FJ, Stroud JL, McGrath SP, Feldmann J, Price AH, Meharg AA (2010) Arsenic shoot-grain relationships in field grown rice cultivars. Environ Sci Technol 44:1471–1477

Odanaka Y, Tsuchiya N, Matano O, Goto S (1985) Characterization of arsenic metabolites in rice plant treated with DSMA (disodium methanearsonate). J Agric Food Chem 33:757–763

Panaullah GM, Alam T, Hossain MB, Loeppert RH, Lauren JG, Meisner CA, Ahmed ZU, Duxbury JM (2009) Arsenic toxicity to rice (*Oryza sativa* L.) in Bangladesh. Plant Soil 317:31–39

Paszkowski U, Kroken S, Roux C, Briggs SP (2002) Rice phosphate transporters include an evolutionarily divergent gene specifically activated in arbuscular mycorrhizal symbiosis. Proc Natl Acad Sci USA 99:13324–13329

Pickering IJ, Prince RC, George MJ, Smith RD, George GN, Salt DE (2000) Reduction and coordination of arsenic in Indian mustard. Plant Physiol 122:1171–1177

Pillai TR, Yan WG, Agrama HA, James WD, Ibrahim AMH, McClung AM, Gentry TJ, Loeppert RH (2010) Total grain-arsenic and arsenic-species concentrations in diverse rice cultivars under flooded conditions. Crop Sci 50:2065–2075

Quaghebeur M, Rengel Z (2003) The distribution of arsenate and arsenite in shoots and roots of *Holcus lanatus* is influenced by arsenic tolerance and arsenate and phosphate supply. Plant Physiol 132:1600–1609

Raab A, Feldmann J, Meharg AA (2004a) The nature of arsenic-phytochelatin complexes in *Holcus lanatus* and *Pteris cretica*. Plant Physiol 134:1113–1122

Raab A, Meharg AA, Jaspars M, Genney DR, Feldmann J (2004b) Arsenic-glutathione complexes – their stability in solution and during separation by different HPLC modes. J Anal Atom Spectrom 19:183–190

Raab A, Schat H, Meharg AA, Feldmann J (2005) Uptake, translocation and transformation of arsenate and arsenite in sunflower (*Helianthus annuus*): formation of arsenic-phytochelatin complexes during exposure to high arsenic concentrations. New Phytol 168:551–558

Raab A, Ferreira K, Meharg AA, Feldmann J (2007a) Can arsenic-phytochelatin complex formation be used as an indicator for toxicity in *Helianthus annuus*? J Exp Bot 58:1333–1338

Raab A, Williams PN, Meharg A, Feldmann J (2007b) Uptake and translocation of inorganic and methylated arsenic species by plants. Environ Chem 4:197–203

Raghothama KG (1999) Phosphate acquisition. Ann Rev Plant Physiol Plant Mol Biol 50:665–693

Rahman MA, Hasegawa H, Rahman MM, Miah MAM, Tasmin A (2008) Straighthead disease of rice (*Oryza sativa* L.) induced by arsenic toxicity. Environ Exp Bot 62:54–59

Rasamivelona A, Gravois KA, Dilday RH (1995) Heritability and genotype x environment interactions for straighthead in rice. Crop Sci 35:1365–1368

Reed JF, Sturgis MB (1936) Toxicity from arsenic compounds to rice on flooded soil. J Am Soc Agron 28:432–436

Requejo R, Tena M (2005) Proteome analysis of maize roots reveals that oxidative stress is a main contributing factor to plant arsenic toxicity. Phytochemistry 66:1519–1528

Schmöger MEV, Oven M, Grill E (2000) Detoxification of arsenic by phytochelatins in plants. Plant Physiol 122:793–801

Sentenac H, Grignon C (1985) Effect of pH on orthophosphate uptake by corn roots. Plant Physiol 77:136–141

Seyfferth AL, Webb SM, Andrews JC, Fendorf S (2010) Arsenic localization, speciation, and co-occurrence with iron on rice (*Oryza sativa* L.) roots having variable Fe coatings. Environ Sci Technol 44:8108–8113

Shin H, Shin HS, Dewbre GR, Harrison MJ (2004) Phosphate transport in *Arabidopsis*: Pht1;1 and Pht1;4 play a major role in phosphate acquisition from both low- and high-phosphate environments. Plant J 39:629–642

Sneller FEC, Van Heerwaarden LM, Kraaijeveld-Smit FJL, Ten Bookum WM, Koevoets PLM, Schat H, Verkleij JAC (1999) Toxicity of arsenate in *Silene vulgaris*, accumulation and degradation of arsenate-induced phytochelatins. New Phytol 144:223–232

Song WY, Park J, Mendoza-Cozatl DG, Suter-Grotemeyer M, Shim D, Hortensteiner S, Geisler M, Weder B, Rea PA, Rentsch D, Schroeder JI, Lee Y, Martinoia E (2010) Arsenic tolerance in Arabidopsis is mediated by two ABCC-type phytochelatin transporters. Proc Natl Acad Sci USA 107:21187–21192

Stroud JL, Norton GJ, Islam MR, Dasgupta T, White R, Price AH, Meharg AA, McGrath SP, Zhao FJ (2011) The dynamics of arsenic in four paddy fields in the Bengal delta. Environ Pollut 159:947–953

Su YH, McGrath SP, Zhao FJ (2010) Rice is more efficient in arsenite uptake and translocation than wheat and barley. Plant Soil 328:27–34

Sun GX, Williams PN, Carey AM, Zhu YG, Deacon C, Raab A, Feldmann J, Islam RM, Meharg AA (2008) Inorganic arsenic in rice bran and its products are an order of magnitude higher than in bulk grain. Environ Sci Technol 42:7542–7546

Takahashi Y, Minamikawa R, Hattori KH, Kurishima K, Kihou N, Yuita K (2004) Arsenic behavior in paddy fields during the cycle of flooded and non-flooded periods. Environ Sci Technol 38:1038–1044

Takamatsu T, Aoki H, Yoshida T (1982) Determination of arsenate, arsenite, monomethylarsonate, and dimethylarsinate in soil polluted with arsenic. Soil Sci 133:239–246

Ullrich-Eberius CI, Sanz A, Novacky AJ (1989) Evaluation of arsenate- and vanadate-associated changes of electrical membrane potential and phosphate transport in *Lemna gibba*-G1. J Exp Bot 40:119–128

Wallace IS, Choi WG, Roberts DM (2006) The structure, function and regulation of the nodulin 26-like intrinsic protein family of plant aquaglyceroporins. Biochim Biophys Acta 1758:1165–1175

Wang JR, Zhao FJ, Meharg AA, Raab A, Feldmann J, McGrath SP (2002) Mechanisms of arsenic hyperaccumulation in *Pteris vittata*. Uptake kinetics, interactions with phosphate, and arsenic speciation. Plant Physiol 130:1552–1561

Wells BR, Gilmour JT (1977) Sterility in rice cultivars as influenced by MSMA rate and water management. Agron J 69:451–454

Williams PN, Price AH, Raab A, Hossain SA, Feldmann J, Meharg AA (2005) Variation in arsenic speciation and concentration in paddy rice related to dietary exposure. Environ Sci Technol 39:5531–5540

Williams PN, Raab A, Feldmann J, Meharg AA (2007a) Market basket survey shows elevated levels of As in South Central US processed rice compared to California: consequences for human dietary exposure. Environ Sci Technol 41:2178–2183

Williams PN, Villada A, Deacon C, Raab A, Figuerola J, Green AJ, Feldmann J, Meharg AA (2007b) Greatly enhanced arsenic shoot assimilation in rice leads to elevated grain levels compared to wheat and barley. Environ Sci Technol 41:6854–6859

Williams PN, Zhang H, Davison W, Meharg AA, Hossain M, Norton GJ, Brammer H, Islam MR (2011) Organic matter-solid phase interactions are critical for predicting arsenic release and plant uptake in Bangladesh paddy soils. Environ Sci Technol 45:6080–6087

Woolson EA, Aharonson N, Iadevaia R (1982) Application of the high-performance liquid-chromatography flameless atomic-absorption method to the study of alkyl arsenical herbicide metabolism in soil. J Agric Food Chem 30:580–584

Wu JH, Zhang R, Lilley RM (2002) Methylation of arsenic in vitro by cell extracts from bentgrass (*Agrostis tenuis*): effect of acute exposure of plants to arsenate. Func Plant Biol 29:73–80

Wu ZC, Ren HY, McGrath SP, Wu P, Zhao FJ (2011) Investigating the contribution of the phosphate transport pathway to arsenic accumulation in rice. Plant Physiol 157:498–508

Xu XY, McGrath SP, Zhao FJ (2007) Rapid reduction of arsenate in the medium mediated by plant roots. New Phytol 176:590–599

Xu XY, McGrath SP, Meharg A, Zhao FJ (2008) Growing rice aerobically markedly decreases arsenic accumulation. Environ Sci Technol 42:5574–5579

Yan WG, Dilday RH, Tai TH, Gibbons JW, McNew RW, Rutger JN (2005) Differential response of rice germplasm to straighthead induced by arsenic. Crop Sci 45:1223–1228

Ye WL, Wood BA, Stroud JL, Andralojc PJ, Raab A, McGrath SP, Feldmann J, Zhao FJ (2010) Arsenic speciation in phloem and xylem exudates of castor bean. Plant Physiol 154:1505–1513

Zavala YJ, Gerads R, Gürleyük H, Duxbury JM (2008) Arsenic in rice: II. Arsenic speciation in USA grain and implications for human health. Environ Sci Technol 42:3861–3866

Zhang H, Selim HM (2008) Reaction and transport of arsenic in soils: equilibrium and kinetic modeling. Adv Agron 98:45–115

Zhao FJ, Wang JR, Barker JHA, Schat H, Bleeker PM, McGrath SP (2003) The role of phytochelatins in arsenic tolerance in the hyperaccumulator *Pteris vittata*. New Phytol 159:403–410

Zhao FJ, Ma JF, Meharg AA, McGrath SP (2009) Arsenic uptake and metabolism in plants. New Phytol 181:777–794

Zhao FJ, Ago Y, Mitani N, Li RY, Su YH, Yamaji N, McGrath SP, Ma JF (2010a) The role of the rice aquaporin Lsi1 in arsenite efflux from roots. New Phytol 186:392–399

Zhao FJ, McGrath SP, Meharg AA (2010b) Arsenic as a food-chain contaminant: mechanisms of plant uptake and metabolism and mitigation strategies. Ann Rev Plant Biol 61:535–559

Zhao FJ, Stroud JL, Khan MA, McGrath SP (2012) Arsenic translocation in rice investigated using radioactive ^{73}As tracer. Plant Soil 350:413–420

Zheng MZ, Cai C, Hu Y, Sun GX, Williams PN, Cui HJ, Li G, Zhao FJ, Zhu YG (2011) Spatial distribution of arsenic and temporal variation of its concentration in rice. New Phytol 189:200–209

Chapter 7
Strategies for Producing Low Arsenic Rice

7.1 Introduction

This chapter will discuss possible strategies to minimize arsenic accumulation in rice. Where grain arsenic concentration is elevated due to ongoing contamination, the ideal scenario is to stop the contamination at the source. However, this may not be practical due to other considerations. For example, there may be no alternative to irrigation with As-containing groundwater during the dry season of Boro rice production in some regions of South Asia. Even when arsenic inputs can be stopped or minimized, the existing level of arsenic in the soil may continue to cause elevated transfer to the food chain. Also, natural geochemical weathering of parent material has given rise to elevated arsenic in grain in many rice growing regions. Therefore, practical methods are needed to deal with this problem.

7.2 Paddy Field Water Management

As discussed in Chap. 5, the reductive mobilization of arsenic under the anaerobic conditions in paddy soils greatly enhances the bioavailability of arsenic to rice, leading to excessive arsenic accumulation in rice grain and straw. In comparison, soils maintained at aerobic conditions generally have very low levels of arsenic in the soil solution. In greenhouse experiments, maintaining soil under aerobic conditions decreased arsenic concentration in rice grain and straw by 10–20, and 7–63 fold, respectively, compared with continuously flooded rice (Li et al. 2009b; Xu et al. 2008). Li et al. (2009b) further investigated the effect of imposing a period of aerobic conditions during either the vegetative or reproductive growth, and showed that these treatments decreased grain arsenic concentration by 80% and 50%, respectively (Fig. 7.1). The pot study by Arao et al. (2009) further supports the results described above, showing that an aerobic treatment 3 weeks before and after

A.A. Meharg and F.J. Zhao, *Arsenic & Rice*, DOI 10.1007/978-94-007-2947-6_7,

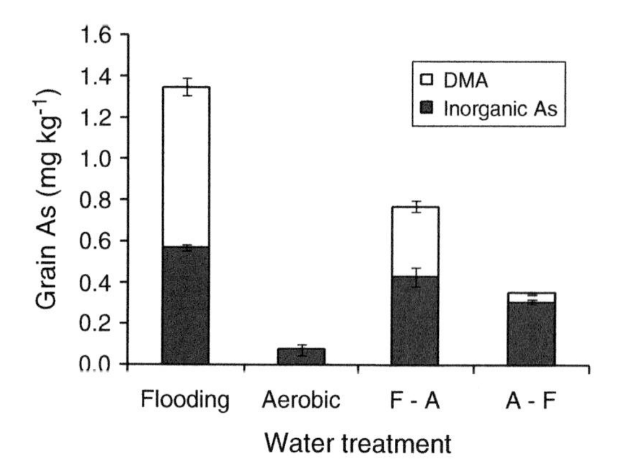

Fig. 7.1 Effect of water management on arsenic concentration in rice grain in a pot experiment. Rice plants were grown under the conditions of flooded, aerobic, flooded followed by aerobic after flowering (*F–A*), or aerobic followed flooded after flowering (*A–F*) (Redrawn with permission from Li et al. (2009b). Copyright (2009) American Chemical Society)

heading was effective in reducing grain arsenic concentration. These greenhouse experiments provide proof of the concept that water management can be a highly effective tool in controlling arsenic bioavailability in paddy soils and the subsequent accumulation in rice grain. Further evidence comes from a field trial at Beaumont, Texas, USA, comparing the performance of more than 370 rice cultivars under flooded or non-flooded conditions. There was a 14-fold difference in the mean arsenic concentration in rice grain between the flooded rice (mean 632 $\mu g\ kg^{-1}$) and the non-flooded rice (mean 45 $\mu g\ kg^{-1}$); the respective ranges were 172–1,682 and 9–126 $\mu g\ kg^{-1}$ (Norton et al. 2012).

Clearly, the problem of excessive arsenic accumulation in rice would greatly diminish if rice could be grown aerobically. However, rice is typically grown under flooded conditions for a reason, and that is because cultivated rice was domesticated from a semi-aquatic perennial ancestor and thus is very sensitive to water shortage (Bouman et al. 2007). Moreover, flooding soil helps control the build-up of pathogens, nematodes and weeds, and can also increase nutrient availability (e.g. phosphorus) (Peng et al. 2006; Ventura et al. 1981). However, conventional cultivation of paddy rice is wasteful in the use of water, which has increasingly become a scarce resource in many rice growing regions. New cultivation methods, such as aerobic rice, alternate wetting and drying (AWD) and raised bed cultivation, have been or are being developed to save water (Bouman et al. 2007). These cultivation methods may also prove to be highly effective in combating the problem of excessive accumulation of arsenic in rice, not only because water-saving methods are likely to maintain soil under more oxic conditions and hence less arsenic mobilization, but also saving water means less input of arsenic into the paddy field from irrigation of As-containing groundwater where elevated arsenic in groundwater exacerbates grain arsenic levels. These methods of cultivation and water management are described in details by Bouman et al. (2007) and a brief introduction is given below.

7.2.1 Aerobic Rice

Aerobic rice is a production system in which "aerobic rice" varieties are grown in well-drained, non-puddled, and nonsaturated soils (Bouman et al. 2007). When rainfall is insufficient, irrigation is applied to bring the soil water content in the root zone up to field capacity after it has reached a certain lower threshold level. Water use is reduced substantially because seepage and percolation are minimized and evaporation is also decreased greatly since there is no ponded water layer (Bouman et al. 2005, 2007). Lowland paddy rice varieties are not suitable for aerobic cultivation because of their sensitivity to drought. Peng et al. (2006) showed that, using the best available aerobic rice varieties in the tropics available at the time, yields of aerobic rice were 8–69% lower than flooded rice. In recent years, cultivars adapted to aerobic conditions have been developed. In northern China, breeders have produced temperate aerobic rice varieties with a yield potential of 6 tonnes per hectare consuming only 50% of the water used in lowland rice (http://irri.org/our-science/aerobic-rice). Breeding of aerobic rice varieties for the tropical region is also under way. Because soil moisture is maintained at nonsaturated conditions, arsenic mobilization is expected to be greatly curtailed, leading to decreased arsenic uptake by aerobic rice. Field trials should be conducted to establish the extent of reduction in arsenic accumulation in aerobic rice, and to evaluate any potential undesirable effects on the accumulation of essential and other toxic elements (see Sect. 7.2.4).

7.2.2 Alternate Wetting and Drying (AWD)

As the name indicates, the water status in paddy field oscillates between wet and dry in the AWD method. Irrigation water is applied to obtain flooded conditions, then the ponded water is allowed to disappear for a number of days before irrigation water is applied again (Bouman et al. 2007). The number of days when soil is not flooded varies from 1 to more than 10 days. A practical way to implement AWD is to monitor the water depth in the field using the "field water tube". When the water level in the tube is 15 cm below the soil surface, it is time to irrigate and flood the soil with a depth of around 5 cm (Bouman et al. 2007). Also, 5-cm depth of ponded water should be maintained during the flowering time (from 1 week before to 1 week after the peak of flowering) to avoid any water stress that would result in potentially severe yield loss. When the 15-cm threshold is used, the method is called "Safe AWD" as it generally does not cause yield losses because the roots of the rice plants are still able to take up water from the saturated soil and the perched water in the root zone. The safe AWD method can save water use by about 15% (Bouman et al. 2007). Larger saving of water is possible by lowering the threshold level for irrigation in the field water tube to 20, 25, 30 cm, or even deeper, although there may be yield penalty.

The effect of adopting AWD on arsenic uptake by rice has not been tested under field conditions. In a paddy field in Bangladesh irrigated intermittently, the arsenic concentration in the soil pore water from the plough layer remained relatively low during the grain filling stage, corresponding to the oxic conditions observed in the soil layer (Roberts et al. 2011). In another field study in Bangladesh, the site employing intermittent irrigation (not strictly AWD) produced lower grain arsenic than the other site under continuously flooded conditions (Stroud et al. 2011). A recent field study at Stuttgart, Arkansas, USA showed that grain contained 41% lower arsenic in intermittently flooded paddy than in continuously flooded paddy (Somenahally et al. 2011).

7.2.3 Raised Bed Cultivation

In this cultivation method, rice is grown on beds that are separated by furrows through which irrigation water is introduced (Bouman et al. 2007). Irrigation is intermittent and the soil in the raised beds is dominantly in aerobic conditions; hence, the method can be considered an aerobic rice system. In general, furrow irrigation is more water efficient than whole field flash flooding. Though dimensions may vary, beds are usually around 35 cm wide, separated by furrows that are 30 cm wide and 25 cm deep (Bouman et al. 2007). Rice can be transplanted or direct-seeded on the beds. So far, the raised-bed system has mostly been tested with current lowland rice varieties, and yield gains can be expected when suitable aerobic varieties are developed.

Duxbury and Panaullah (2007) grew Boro rice in paired plots employing either conventional flooded cultivation or raised bed cultivation. The plots were established across a soil arsenic gradient from 12 to 58 mg kg^{-1} as a result of past irrigation of As-contaminated tube-well water in the Faridpur district, Bangladesh. Under the conventional flooded paddy conditions, grain yield decreased more or less linearly from 8.9 to 3.0 t ha^{-1} across the soil arsenic gradient due to arsenic toxicity. In contrast, rice yield was less affected by increasing soil arsenic under the raised bed cultivation, declining from 8.0 to 5.2 t ha^{-1}. Soil redox potential in the top 15 cm layer was about 0 and −120 mV in the raised bed and conventional systems, respectively. Cultivation of rice on the raised beds also decreased the arsenic concentrations of rice grain and straw compared with the conventional flooded conditions. At the two lower levels of soil arsenic contamination (12 and 26 mg kg^{-1}), grain arsenic concentration was approximately halved and straw arsenic was decreased by sevenfold by the raised bed cultivation. At the two higher levels of soil arsenic contamination (40 and 58 mg kg^{-1}), there was still a threefold difference in straw arsenic concentration between the two cultivation methods, but the concentration of arsenic in grain became comparable.

Talukder et al. (2011) compared raised bed with conventional flooded cultivation methods in an arsenic affected area of Bangladesh. The soil contained 8.1 mg kg^{-1} total arsenic and the irrigation water 0.10 mg As L^{-1}. The raised bed system saved water use by about 30% compared with the conventional method and maintained the soil redox potential above that of the arsenate/arsenite redox half reaction (Table 7.1). There was a 13% increase in grain yield over the conventional cultivation method

Table 7.1 Effect of raised bed cultivation on grain yield, water use, and arsenic concentrations of grain and straw

Cultivation method	Water use (mm)		Soil redox potential (mV)	Grain yield (t ha^{-1})		As concentration (mg kg^{-1})	
	Boro	Aman		Boro	Aman	Grain	Straw
Conventional	1,187	268	−56	6.06	5.61	0.65	10.9
Raised bed	827	192	+361	6.88	6.38	0.25	1.55

Adapted from Talukder et al. (2011)
Note: data are means of two seasons (2004 and 2005) except for soil redox potential which was measured in 2005 only (mean of the 10–30 cm depth). Soil pH was similar for the two cultivation methods (6.6–6.7). Grain yield was for the normal rate of phosphorus application

Table 7.2 Effect of water management on the concentrations (mg kg^{-1}) of Cd, Fe, Zn and Se in rice grain in a pot experiment

Water regime	Cd	Fe	Zn	Se
Flooded	0.03	16.4	24.4	0.23
Aerobic	0.26	13.6	43.8	0.06
Flooded-aerobic	0.11	11.9	25.2	0.37
Aerobic-flooded	0.15	16.8	38.6	0.26

Unpublished data from the experiment of Li et al. (2009b)

and, importantly, arsenic concentrations in grain and straw were decreased by 62% and 86%, respectively (Table 7.1). These results demonstrate the considerable benefits of the raised bed cultivation in water saving, yield increase and reduction in arsenic uptake.

7.2.4 *Effects of Water Management on the Accumulation of Other Elements*

Although arsenic accumulation in rice is a prominent issue, accumulation of essential micronutrients and other toxic elements should also be considered. Polished rice is low in the concentrations of the essential micronutrients iron and zinc; deficiencies in these micronutrients affect a large number of people worldwide (White and Broadley 2009; Zhao and Shewry 2011). Rice produced in many regions of the world is also low in the selenium concentration (Williams et al. 2009). On the other hand, rice is a major source of the toxic trace element cadmium for the population based on a rice staple diet (Arao et al. 2009; Ueno et al. 2010).

Whilst aerobic cultivation of rice decreases arsenic accumulation, it can lead to excessive accumulation of cadmium (Arao et al. 2009). In the pot study of Li et al. (2009b), aerobic treatment during the entire growth period increased grain cadmium concentration by ninefold compared with the continuously flooded treatment (Table 7.2), although grain cadmium concentration was still below the limit (0.4 mg kg^{-1}) for polished rice set by the Codex Alimentarius Commission of the

Food and Agriculture Organization (2006). For the aerobic followed by flooded or flooded followed by aerobic treatments, grain cadmium was intermediate between the continuously flooded and continuously aerobic treatments. It is well known that anaerobic soil conditions decrease cadmium bioavailability, but the exact mechanisms are not entirely understood. Possible explanations include the precipitations of cadmium sulphide or cadmium carbonate in soil under anaerobic conditions, or that Fe^{2+} ions mobilized in flooded soil suppress Cd^{2+} uptake by plants (de Livera et al. 2011; Khaokaew et al. 2011). Other effects of aerobic treatment include decreased concentrations of selenium and iron, but increased concentration of zinc (Table 7.2). Therefore, when considering the use of aerobic cultivation to control arsenic accumulation, other effects should also be taken into account. Where the soil cadmium level is high, flooding soil is an effective way of decreasing its uptake by rice.

7.3 Cultivar Selection and Breeding Low Arsenic Rice

One way to reduce arsenic accumulation in rice grain is to identify the extent of genetic variation in grain arsenic concentration and speciation. Low arsenic cultivars that are adapted to the local environment can be grown selectively, especially in the areas where arsenic contamination is a problem. Furthermore, low arsenic cultivars can be used as genetic stock for further breeding purpose.

7.3.1 Cultivar Difference in Arsenic Accumulation

A number of recent field studies have shown substantial genetic variation in grain arsenic concentration as well as arsenic speciation. Field trials at Faridpur and Sonargaon, Bangladesh, showed 4–4.6 fold variation in total grain arsenic among 76 cultivars including local landraces, locally improved cultivars and parents of permanent mapping populations (Norton et al. 2009b). Local landraces with red bran had the highest arsenic concentration, although it is not known whether there exists a genetic linkage between bran colour and arsenic accumulation. The two field sites had both been impacted by As-contaminated groundwater, but had different levels of arsenic in the soils and in the irrigation water. Despite the site differences, there was a strong correlation ($r=0.802$) in grain arsenic concentrations of the 76 cultivars between the two sites, suggesting stable genetic differences in arsenic accumulation, at least at the two experimental locations (Norton et al. 2009b). When comparisons were made across six sites in three countries (two each in Bangladesh, West Bengal, India and south China), the cultivar rankings were broadly consistent between the two Bangladeshi sites and between the two Indian sites, but not between the two Chinese sites

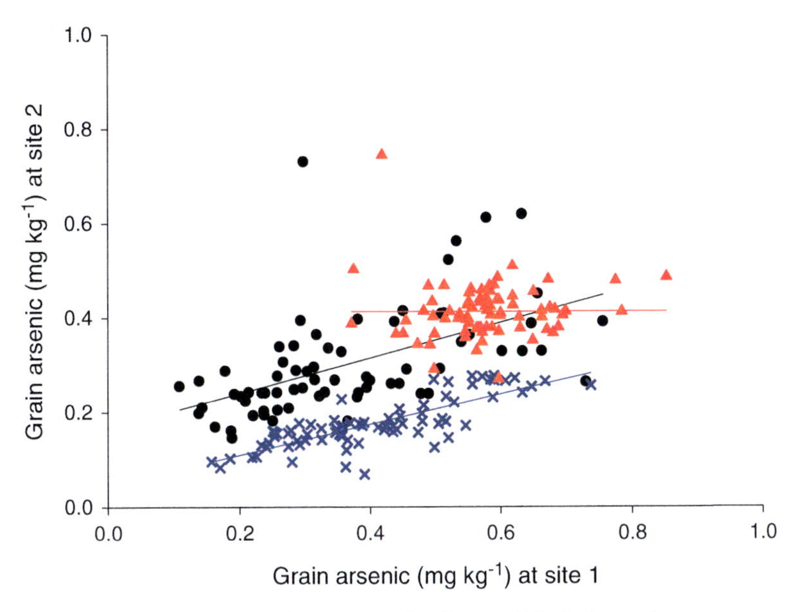

Fig. 7.2 Correlation of total arsenic concentration in unpolished rice grain among common cultivars at six field sites in three different countries. *Red triangles* represent Chinese field sites, *Blues crosses* represent Bangladeshi field sites and *black circles* represent Indian field sites. For each country, sites with the highest overall arsenic were designated as site 1 (Reprinted with permission from Norton et al. (2009a). Copyright (2009) American Chemical Society)

(Fig. 7.2). Among the 13 common cultivars grown at all six sites, environment made the largest contribution to the variation in grain arsenic concentration (61%), followed by genotype (6%) and genotype x environment interaction (19%) (Norton et al. 2009a). These results indicate a genotype x environment interaction across diverse environments, which is not surprising considering that arsenic bioavailability in soil is greatly influenced by soil properties, the source of arsenic contamination, water management and other environmental factors (see Chap. 5). In the studies of Norton et al. (2009a, b), there was a significant genotype effect on the percentages of inorganic arsenic and DMA in grain, but the environmental influence was greater. All except one site in China showed a positive relationship between total grain arsenic and the percentage of DMA, and a negative relationship between total grain arsenic and the percentage of inorganic arsenic. These results emphasize the environmental impact on arsenic speciation in rice grain.

Similarly, Ahmed et al. (2011) reported a large environmental effect in their trials of 38 Bangladeshi cultivars grown at ten experimental sites across different agro-ecological zones of Bangladesh. Environment accounted for 69–80% of the variation in grain arsenic concentration, whereas genotype and genotype x environment interactions accounted for only 9–10% and 10–21% of the observed variability. The dominance of environment over genetic in the study of Ahmed et al. (2011) is not unexpected because the cultivars tested represented rather narrow genetic diversity and the environments of the experimental sites were diverse. Nevertheless, both the studies of Norton et al. (2009a, b) and Ahmed et al. (2011) identified a number of rice cultivars that accumulated relatively low levels of arsenic across sites.

Pillai et al. (2010) reported a study at a single field site near Stuttgart, Arkansas, USA, over three growing seasons. They found that the concentrations of total grain arsenic and arsenic species (arsenite and DMA) varied widely among 25 diverse rice cultivars. Arsenic concentration and speciation were mostly dependent on genotype, which accounted for about 70% of the variation in total grain arsenic. The effects of year and genotype by year interactions were also significant, but together accounted for about 20% of the variation. Comparing the studies of Pillai et al. (2010) with those of Ahmed et al. (2011) and Norton et al. (2009a), it may be concluded that genetic stability is greater across seasons than across diverse sites. Interestingly, Pillai et al. (2010) reported a significant positive correlation between the concentration of DMA in grain and the number of days to heading. There was also a significant positive correlation between grain arsenite concentration and days to heading in one season. It is possible that a longer period of vegetative growth (i.e. later flowering) leads to prolonged exposure of the plants to arsenite and methylated arsenic species in flooded soil.

In a study of ten major rice cultivars grown in Japan, the variability in total grain arsenic was significant but relatively small (about 1.6-fold) (Kuramata et al. 2011). However, rice grown in pots with a high-As soil produced on average 17-fold higher grain arsenic than the same cultivars grown in a paddy field with a low-As soil; the difference was mainly attributed to much larger concentrations of DMA in the pot-grown rice. Even when soil and rice cultivar were the same, pot-grown rice grain contained larger percentages of DMA than field-grown rice, possibly because soil was maintained under continuously flooded and stagnant conditions in pots, whereas in the field soil redox potential may be periodically higher due to the dynamics of water movement and occasional draining (Khan et al. 2010).

Accumulation of arsenic in rice grain involves a number of steps and checkpoints which control the uptake and translocation of arsenic from roots to grain (see Chap. 6), giving rise to potential differences among genotypes. Rhizosphere characteristics may also play a role. Pot studies by Mei et al. (2009) and Wu et al. (2011a) showed significant negative correlations between arsenic concentrations in straw or grain with root porosity or radial oxygen loss from the roots among 20–25 rice cultivars. The mechanism underlying these negative correlations remains unclear, but it can be envisaged that cultivars able to release more oxygen may maintain the rhizosphere at a higher redox potential and form more iron plaque on the root surface, which may reduce arsenic uptake. Oxygen released into the rhizosphere may enhance the oxidation of arsenite to arsenate; the latter is more strongly adsorbed by the iron plaque.

7.3.2 *Quantitative Trait Loci for Arsenic Accumulation*

Zhang et al. (2008) identified two quantitative trait loci (QTLs) for arsenic concentration in grain in a doubled-haploid population between a *Japonica* and an *Indica* cultivar grown in pots. The two QTLs are located on chromosomes 6 and 8 and explained 26% and 35% of the phenotypic variance, respectively. One QTL each was also found for shoot and root arsenic concentrations at the seedling stage. These are located on chromosomes 2 and 3, respectively, which are different from those for grain arsenic concentration. In a field experiment with a F_6 recombinant inbred line population derived from a cross between Bala (*Indica*) and Azucena (*Japonica*), Norton et al. (2010) found no main effect QTL for grain As, but there was a significant epistatic interaction between the bottom regions of chromosomes 1 and 2. There were five QTLs for leaf arsenic concentration, each explaining a relatively small proportion (11–18%) of the phenotypic variance. The QTLs reported by Zhang et al. (2008) and Norton et al. (2010) are different, either because of the different mapping populations used or because environmental conditions exert a strong influence on arsenic accumulation or both. Robust QTLs are those that are consistent across different environments, which, once identified, are particularly useful for the development of new cultivars with low arsenic accumulation through marker-assisted breeding.

7.4 Fertilization and Soil Amendments

In a pot experiment, addition of silicon gel (SiO_2) to soil decreased arsenic concentrations in straw and grain by 78% and 16%, respectively (Fig. 7.3) (Li et al. 2009b). These effects occurred even though Si increased soluble arsenite and arsenate in the soil solution, possibly through ligand exchange with the adsorbed arsenic species (see Chap. 5). As discussed in Chap. 6, the effect of Si is due to the competition of silicic acid with arsenite for the transporter Lsi2 in rice roots. In the same experiment, Si addition decreased the concentration of inorganic arsenic in grain by about 60% but increased the DMA concentration by 33% (Fig. 7.3) (Li et al. 2009b). The effect on DMA is attributed to the soil chemical process; Si addition increased the DMA concentration in soil solution (Liu and Zhao, unpublished data). Although DMA, like arsenite, is taken up by rice roots partly through the Lsi1 aquaporin channel, this transport process appears to be subject to little competition from silicic acid, perhaps because the flux through aquaporin channels is very fast (Li et al. 2009a). For arsenite uptake, the competition occurs during the transport mediated by Lsi2, which is a carrier protein. Because Lsi2 does not transport MMA or DMA (Li et al. 2009a), there is no competitive effect of silicic acid on the accumulation of MMA or DMA in rice shoots, unlike for arsenite.

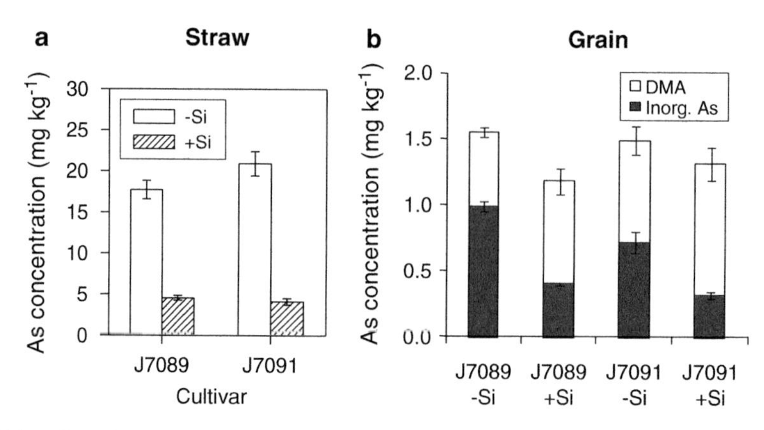

Fig. 7.3 Effect of silicon fertilizer on arsenic concentration in straw and arsenic speciation in grain of two rice cultivars in a pot experiment (Redrawn with permission from Li et al. (2009b). Copyright (2009) American Chemical Society)

Many forms of silicon fertilizers are available, some of which are industrial byproducts, such as basic slag from the steel manufacture processes (Ma and Takahashi 2002). Silicon fertilizers have been widely used in Japan to increase rice yield, but are rarely used in South Asian countries such as Bangladesh. Whether silicon fertilization can effectively decrease arsenic accumulation in rice in the areas affected by arsenic contamination should be evaluated under field conditions, taking into consideration the effects on yield and arsenic accumulation as well as the cost and availability of the fertilizers. It should be noted that some silicon fertilizers are alkaline; their use could increase soil pH leading to increased partition of adsorbed arsenic to the soil solution phase (see Chap. 5), which may outweigh the competitive effect of silicic acid on arsenite uptake by rice.

A number of pot and field experiments have shown that additions of phosphate fertilizer to soil did not decrease arsenic uptake by rice; in fact, an opposite effect was observed in some cases (Abedin et al. 2002; Hossain et al. 2009; Talukder et al. 2011; Wu et al. 2011b). Phosphate is effective in exchanging adsorbed arsenate or arsenite from the soil solid phase and from the iron plaque on the root surface, thus increasing the arsenic availability to rice plants. This effect may be amplified if arsenate is reduced to arsenite in soil solution, which is then taken up by rice roots through non-phosphate transport pathways. Indeed, a recent study showed that the phosphate transport pathway made little contribution to the arsenic accumulation in rice plants grown in flooded soil (Wu et al. 2011b). A lesson from these studies is that excessive use of phosphate fertilizers should be avoided.

Iron oxides/hydroxides are an important component in the soil–paddy rice system with a strong influence on the arsenic biogeochemistry (see Chap. 5). Hossain et al. (2009) tested the effect of adding ferrous sulphate on arsenic toxicity and uptake by rice in a pot experiment. When arsenic was added to the soil, an addition of ferrous sulphate alleviated arsenic toxicity in rice plants, possibly by enhancing the formation of iron plaque on the root surface. However, the effect on arsenic accumulation in straw and grain was mostly not significant. The role of iron plaque is complex, possibly serving both as a sink that concentrates arsenic in the rhizosphere and as a barrier to the entry of arsenic into the root cells (see Chap. 6). Also, use of ferrous iron to treat large areas of paddy soil is unlikely to be practical.

The nutritional status of sulphur can influence arsenic detoxification and mobility within plant due to the role of thio-rich compounds (e.g. phytochelatins) in complexing and sequestering arsenite (Zhao et al. 2010) (see Chap. 6). Plants deficient in sulphur were found to have a greater root-to-shoot translocation of arsenic (Liu et al. 2010; Zhang et al. 2011). Another effect is an apparent increase of iron plaque formation in rice plants treated with sulphur fertilizers and small decreases in the concentrations of arsenic in rice tissues (Hu et al. 2007). Whether applications of sulphur fertilizers can decrease arsenic accumulation in rice grain should be investigated in field trials. Where sulphur deficiency is likely to occur, it does make sense to apply sulphur fertilizers to prevent yield losses and potential elevation of arsenic accumulation.

References

Abedin MJ, Cresser MS, Meharg AA, Feldmann J, Cotter-Howells J (2002) Arsenic accumulation and metabolism in rice (*Oryza sativa* L.). Environ Sci Technol 36:962–968

Ahmed ZU, Panaullah GM, Gauch H, McCouch SR, Tyagi W, Kabir MS, Duxbury JM (2011) Genotype and environment effects on rice (*Oryza sativa* L.) grain arsenic concentration in Bangladesh. Plant Soil 338:367–382

Arao T, Kawasaki A, Baba K, Mori S, Matsumoto S (2009) Effects of water management on cadmium and arsenic accumulation and dimethylarsinic acid concentrations in Japanese rice. Environ Sci Technol 43:9361–9367

Bouman BAM, Peng S, Castaneda AR, Visperas RM (2005) Yield and water use of irrigated tropical aerobic rice systems. Agric Water Manag 74:87–105

Bouman BAM, Lampayan RM, Tuong TP (2007) Water management in irrigated rice: coping with water scarcity. International Rice Research Institute, Los Baños, 54p

Codex Alimentarius Commission (2006) Report of the 29th session of the Codex Alimentarius Commission, ALINORM 06/29/41. Codex Alimentarius Commission, FAO, Rome

de Livera J, McLaughlin MJ, Hettiarachchi GM, Kirby JK, Beak DG (2011) Cadmium solubility in paddy soils: effects of soil oxidation, metal sulfides and competitive ions. Sci Total Environ 409:1489–1497

Duxbury JM, Panaullah G (2007) Remediation of arsenic for agriculture sustainability, food security and health in Bangladesh. FAO Water Working Paper. FAO, Rome, 28p.

Hossain MB, Jahiruddin M, Loeppert RH, Panaullah GM, Islam MR, Duxbury JM (2009) The effects of iron plaque and phosphorus on yield and arsenic accumulation in rice. Plant Soil 317:167–176

Hu ZY, Zhu YG, Li M, Zhang LG, Cao ZH, Smith EA (2007) Sulfur (S)-induced enhancement of iron plaque formation in the rhizosphere reduces arsenic accumulation in rice (*Oryza sativa* L.) seedlings. Environ Pollut 147:387–393

Khan MA, Stroud JL, Zhu YG, McGrath SP, Zhao FJ (2010) Arsenic bioavailability to rice is elevated in Bangladeshi paddy soils. Environ Sci Technol 44:8515–8521

Khaokaew S, Chaney RL, Landrot G, Ginder-Vogel M, Sparks DL (2011) Speciation and release kinetics of cadmium in an alkaline paddy soil under various flooding periods and draining conditions. Environ Sci Technol 45:4249–4255

Kuramata M, Abe T, Matsumoto S, Ishikawa S (2011) Arsenic accumulation and speciation in Japanese paddy rice cultivars. Soil Sci Plant Nutr 57:248–258
Li RY, Ago Y, Liu WJ, Mitani N, Feldmann J, McGrath SP, Ma JF, Zhao FJ (2009a) The rice aquaporin Lsi1 mediates uptake of methylated arsenic species. Plant Physiol 150:2071–2080
Li RY, Stroud JL, Ma JF, McGrath SP, Zhao FJ (2009b) Mitigation of arsenic accumulation in rice with water management and silicon fertilization. Environ Sci Technol 43:3778–3783
Liu WJ, Wood BA, Raab A, McGrath SP, Zhao FJ, Feldmann J (2010) Complexation of arsenite with phytochelatins reduces arsenite efflux and translocation from roots to shoots in *Arabidopsis*. Plant Physiol 152:2211–2221
Ma JF, Takahashi E (2002) Soil, fertilizer and plant silicon research in Japan. Elsevier, Amsterdam
Mei XQ, Ye ZH, Wong MH (2009) The relationship of root porosity and radial oxygen loss on arsenic tolerance and uptake in rice grains and straw. Environ Pollut 157:2550–2557
Norton GJ, Duan GL, Dasgupta T, Islam MR, Lei M, Zhu YG, Deacon CM, Moran AC, Islam S, Zhao FJ, Stroud JL, McGrath SP, Feldmann J, Price AH, Meharg AA (2009a) Environmental and genetic control of arsenic accumulation and speciation in rice grain: comparing a range of common cultivars grown in contaminated sites across Bangladesh, China, and India. Environ Sci Technol 43:8381–8386
Norton GJ, Islam MR, Deacon CM, Zhao FJ, Stroud JL, McGrath SP, Islam S, Jahiruddin M, Feldmann J, Price AH, Meharg AA (2009b) Identification of low inorganic and total grain arsenic rice cultivars from Bangladesh. Environ Sci Technol 43:6070–6075
Norton GJ, Deacon CM, Xiong LZ, Huang SY, Meharg AA, Price AH (2010) Genetic mapping of the rice ionome in leaves and grain: identification of QTLs for 17 elements including arsenic, cadmium, iron and selenium. Plant Soil 329:139–153
Norton GJ, Pinson SRM, Alexander J, Mckay S, Hansen H, Duan GL, Islam MR, Islam S, Stroud JL, Zhao FJ, McGrath SP, Zhu YG, Lahner B, Yakubova E, Guerinot ML, Tarpley L, Eizenga GC, Salt DE, Meharg AA, Price AH (2012) Variation in grain arsenic assessed in a diverse panel of rice (Oryza sativa) grown in multiple sites. New Phytol doi:10.1111/j.1469-8137.2011.03983.x
Peng SB, Bouman B, Visperas RA, Castaneda A, Nie LX, Park HK (2006) Comparison between aerobic and flooded rice in the tropics: agronomic performance in an eight-season experiment. Field Crop Res 96:252–259
Pillai TR, Yan WG, Agrama HA, James WD, Ibrahim AMH, McClung AM, Gentry TJ, Loeppert RH (2010) Total grain-arsenic and arsenic-species concentrations in diverse rice cultivars under flooded conditions. Crop Sci 50:2065–2075
Roberts LC, Hug SJ, Voegelin A, Dittmar J, Kretzschmar R, Wehrli B, Saha GC, Badruzzaman ABM, Ali MA (2011) Arsenic dynamics in porewater of an intermittently irrigated paddy field in Bangladesh. Environ Sci Technol 45:971–976
Somenahally AC, Hollister EB, Yan W, Gentry TJ, Loeppert RH (2011) Water management impacts on arsenic speciation and iron-reducing bacteria in contrasting rice-rhizosphere compartments. Environ Sci Technol 45:8328–8335
Stroud JL, Norton GJ, Islam MR, Dasgupta T, White R, Price AH, Meharg AA, McGrath SP, Zhao FJ (2011) The dynamics of arsenic in four paddy fields in the Bengal delta. Environ Pollut 159:947–953
Talukder A, Meisner CA, Sarkar MAR, Islam MS (2011) Effect of water management, tillage options and phosphorus status on arsenic uptake in rice. Ecotoxicol Environ Saf 74:834–839
Ueno D, Yamaji N, Kono I, Huang CF, Ando T, Yano M, Ma JF (2010) Gene limiting cadmium accumulation in rice. Proc Natl Acad Sci U S A 107:16500–16505
Ventura W, Watanabe I, Castillo MB, Delacruz A (1981) Involvement of nematodes in the soil sickness of a dryland rice-based cropping system. Soil Sci Plant Nutr 27:305–315
White PJ, Broadley MR (2009) Biofortification of crops with seven mineral elements often lacking in human diets – iron, zinc, copper, calcium, magnesium, selenium and iodine. New Phytol 182:49–84

Williams PN, Lombi E, Sun GX, Scheckel K, Zhu YG, Feng XB, Zhu JM, Carey AM, Adomako E, Lawgali Y, Deacon C, Meharg AA (2009) Selenium characterization in the global rice supply chain. Environ Sci Technol 43:6024–6030

Wu C, Ye ZH, Shu WS, Zhu YG, Wong MH (2011a) Arsenic accumulation and speciation in rice are affected by root aeration and variation of genotypes. J Exp Bot 62:2889–2898

Wu ZC, Ren HY, McGrath SP, Wu P, Zhao FJ (2011b) Investigating the contribution of the phosphate transport pathway to arsenic accumulation in rice. Plant Physiol 157:498–508

Xu XY, McGrath SP, Meharg A, Zhao FJ (2008) Growing rice aerobically markedly decreases arsenic accumulation. Environ Sci Technol 42:5574–5579

Zhang J, Zhu YG, Zeng DL, Cheng WD, Qian Q, Duan GL (2008) Mapping quantitative trait loci associated with arsenic accumulation in rice (*Oryza sativa*). New Phytol 177:350–355

Zhang J, Zhao QZ, Duan GL, Huang YC (2011) Influence of sulphur on arsenic accumulation and metabolism in rice seedlings. Environ Exp Bot 72:34–40

Zhao FJ, Shewry PR (2011) Recent developments in modifying crops and agronomic practice to improve human health. Food Policy 36:S94–S101

Zhao FJ, McGrath SP, Meharg AA (2010) Arsenic as a food-chain contaminant: mechanisms of plant uptake and metabolism and mitigation strategies. Annu Rev Plant Biol 61:535–559

Chapter 8
Arsenic in Other Crops

8.1 Introduction

The bulk of this book is devoted to the arsenic biogeochemical cycles in the paddy rice systems, the mechanisms of arsenic uptake and accumulation by rice plants, dietary arsenic intake from rice and potential mitigation methods to reduce arsenic accumulation in rice. The reasons for the focus on rice are explained in Chap. 1. Nevertheless, it is necessary to consider other food crops with regard to their arsenic contents, speciation and likely contributions to the dietary intake of As, which is the task of this chapter.

What makes rice so special with respect to arsenic is arsenic mobilization under reduced conditions that prevail in paddies (see Chap. 5), and the highly efficient assimilation of the mobilized arsenite through silicic acid transporters due to the fact that rice is a silicon accumulator (see Chap. 6). The vast majority of the world's crops, including high starch dietary staples (grain, excluding rice, and tuber crops) are grown under aerobic conditions and are less prone to arsenic accumulation, except where crops are grown on highly polluted soils. This chapter will survey arsenic contamination in crops other than rice.

8.2 Wheat

Wheat is one of the three major cereal crops (the other two being maize and rice), with over 600 million tonnes being produced annually (Shewry 2009). Because wheat is grown on aerobic soils, it generally accumulates relatively low levels of arsenic in the grain. This is confirmed by analysis of over 2,000 cereal samples (excluding rice) in the European Union, showing a mean of 15 $\mu g\ kg^{-1}$ (lower bound) – 40 $\mu g\ kg^{-1}$ fresh weight (upper bound) (Table 2.5) (European Food Safety Authority 2009). Other studies have reported a mean arsenic concentration in wheat of <60 $\mu g\ kg^{-1}$ dry weight for samples collected from non-contaminated areas

A.A. Meharg and F.J. Zhao, *Arsenic & Rice*, DOI 10.1007/978-94-007-2947-6_8,

(Table 8.1). Conversion between the concentrations based on fresh weight and dry weight can be done by assuming 85–90% dry matter content typical of wheat grain. Compared with rice (see Chap. 2), the levels of arsenic in wheat are substantially lower. However, where soil is contaminated with arsenic or As-contaminated water is used for irrigation, wheat grain may accumulate elevated levels of arsenic (Table 8.1). Examples include individual wheat grain samples collected in West Bengal, India, containing 362 and 730 μg As kg^{-1} dry weight (Norra et al. 2005; Roychowdhury et al. 2002); the area has been impacted by arsenic contamination from groundwater. Irrigation with As-contaminated groundwater in Pakistan also resulted in high concentrations (220–620 μg As kg^{-1} dry weight) in wheat grain, compared with a mean concentration of 50 μg As kg^{-1} dry weight when low-As canal water was used (Baig et al. 2011). Relatively high arsenic (up to 500 As kg^{-1} dry weight) was also found in some grain samples collected in Cornwall, Southwest England, where some soils are heavily contaminated by arsenic from the past mining and smelting activities (Williams et al. 2007).

Figure 8.1 presents a log-log plot of the arsenic concentration in wheat grain versus the total arsenic concentration in soil, showing a significant and positive correlation between soil and grain arsenic. It is clear from Fig. 8.1 that the majority of wheat grain from uncontaminated soils (e.g. <10 mg As kg^{-1}) contains <50 μg As kg^{-1} dry weight. Therefore, unlike rice which accumulates arsenic to levels of concern even from soils with background levels of arsenic, its accumulation in wheat may present a problem only if the soil, the irrigation water or the manures used are contaminated.

Relatively low bioavailability of arsenic in aerobic soils is the main reason for a limited arsenic accumulation in wheat grain. Wheat plants also appear to be less efficient than rice in the translocation of arsenic from roots to shoots, which can be attributed to the strong silicon/arsenite transport pathway in the latter species (Su et al. 2010). Williams et al. (2007) showed that the median shoot/soil arsenic concentration ratio in wheat was nearly 50 times lower than that in rice. However, once in the shoots, arsenic may have a greater mobility to move to grain in wheat than in rice, as indicated by the larger grain/shoot arsenic concentration ratio in the former (median 0.19 and 0.05 for wheat and rice, respectively) (Williams et al. 2007).

The localisation of arsenic in wheat grain has been investigated by milling. Similar to rice grain, the bran fraction of wheat (comprising mainly pericarp, aleurone and embryo) contains 3.8–4.7-fold higher concentration of arsenic than the white flour (Zhao et al. 2010). More than half (53–65%) of the arsenic in wheat grain was found in the bran, even though this fraction accounted for only 23–29% of the total grain weight (Zhao et al. 2010). Thus, milling of wheat can substantially decrease the arsenic concentration in the white flour.

Arsenic speciation in wheat grain is dominated by inorganic As. Zhao et al. (2010) detected only inorganic arsenic in the extracts of wheat flour, with As(III) being the main species. Cubadda et al. (2010) found the presence of inorganic arsenic, DMA and traces of MMA in Italian wheat. Inorganic arsenic (mainly As(III)) accounted for on average 95% of the total As, with the remainder being methylated arsenic (mainly DMA). Similarly, Reyes et al. (2007) reported 96% and 4% of

Table 8.1 Total As concentrations in cereals, potato, pulses, vegetables and fruits

Crop	Sample type	Country	As contamination	n	Range (μg kg⁻¹)	Mean (μg kg⁻¹)	Unit base	References
Wheat	Grain	The Netherlands		84	5–285	45	FW	Wiersma et al. (1986)
	Wholemeal flour	USA		4		39	FW	Schoof et al. (1999)
	Grain	West Bengal, India	Water, soil	2	4–10	7	DW	Roychowdhury et al. (2002)
	Grain	West Bengal, India	Water, soil	1		352	DW	Roychowdhury et al. (2002)
	Coarse flour	West Bengal, India	Water, soil	2	20–39	30	DW	Roychowdhury et al. (2002)
	Grain	West Bengal, India	Water, soil	3	710–740	730	DW	Norra et al. (2005)
	Grain	West Bengal, India	Water, soil	8	10–190	129	DW	Bhattacharya et al. (2010)
	Grain	Pakistan		40		50	DW	Baig et al. (2011)
	Grain	Pakistan	Water	40	220–620		DW	Baig et al. (2011)
	Grain	Serbia		52		55	DW	Skrbic and Cupic (2005)
	Grain	Scotland, UK		29	10–210	30	DW	Williams et al. (2007)
	Grain	Cornwall, UK	Soil	37	10–500	70	DW	Williams et al. (2007)
	Wholemeal flour	UK		26[a]	0–7	3	DW	Zhao et al. (2010)
	Wholemeal flour	Poland		26[a]	8–20	13	DW	Zhao et al. (2010)
	Wholemeal flour	Hungary		52[b]	2–25	7	DW	Zhao et al. (2010)
	Wholemeal flour	France	Soil	26[a]	41–101	69	DW	Zhao et al. (2010)
	Grain, bread wheat	Italy		80		8	DW	Cubadda et al. (2010)
	Grain, durum wheat	Italy		61		9	DW	Cubadda et al. (2010)
	Grain	China, USA		19	10–220	50	DW	Adomako et al. (2011)

(continued)

Table 8.1 (continued)

Crop	Sample type	Country	As contamination	n	Range (μg kg^{-1})	Mean (μg kg^{-1})	Unit base	References
Maize	Corn meal	US		4		39	FW	Schoof et al. (1999)
	Corn meal	US		29	0–150	30	FW	Jelinek and Corneliussen (1977)
	Grain	US		32	40–60		DW	Liebhardt (1976)
	Grain	Pakistan		55		40	DW	Baig et al. (2011)
	Grain	Pakistan	Water	55	190–390		DW	Baig et al. (2011)
	Grain	Chile	Water	6	15–152		FW	Diaz et al. (2004)
	Grain	Ghana, China, Mexico, USA		89	3–130	10	DW	Adomako et al. (2011)
Potato	Tuber	The Netherlands		94	2–41	13	FW	Wiersma et al. (1986)
	Tuber	USA		268	0–230	8	FW	Jelinek and Corneliussen (1977)
	Tuber, peeled	West Bengal, India	Water, soil	20	<0.04–29	4	DW	Roychowdhury et al. (2002)
	Tuber skin	West Bengal, India	Water, soil	10	59–690	255	DW	Roychowdhury et al. (2002)
	Tuber	West Bengal, India	Water, soil	23	190–1,020	654	DW	Bhattacharya et al. (2010)
	Tuber	Nepal		15		<10	DW	Dahal et al. (2008)
	Tuber	Bangladesh	Water, soil	5	70–1,390		DW	Das et al. (2004)
	Tuber, peeled	Cornwall, UK	Soil	1	100		DW	Warren et al. (2003)
	Tuber skin	Cornwall, UK	Soil	1	350		DW	Warren et al. (2003)
	Tuber	Bangladesh	Water, soil	3	50–890	380	DW	Williams et al. (2006)
	Tuber	West Bengal, India	Water			80	DW	Signes-Pastor et al. (2008)
	Tuber	Chile		3		6	FW	Munoz et al. (2005)

Pulses	Bean	West Bengal, India	Water, soil	8	<0.04–200	44	DW	Roychowdhury et al. (2002)
	Lentil	West Bengal, India	Water, soil	11	<0.04–117	15	DW	Roychowdhury et al. (2002)
	Lentil	West Bengal, India	Water, soil	14	<0.3–140	96	DW	Bhattacharya et al. (2010)
	Mung bean	Bangladesh	Water, soil	5	<40–70	30	DW	Williams et al. (2006)
	Lentil	Bangladesh	Water, soil	5	<40–90	40	DW	Williams et al. (2006)
	Chickpea	Bangladesh	Water, soil	5	<40–90	50	DW	Williams et al. (2006)
	Lentil	Spain		3		29	DW	Matos-Reyes et al. (2010)
	Bean	Spain		9	13–30		DW	Matos-Reyes et al. (2010)
	Chickpea	Spain		3		61	DW	Matos-Reyes et al. (2010)
	Legumes	Chile		3		8	FW	Munoz et al. (2005)
Vegetables	Leafy vegetables	Bangladesh	Water, soil	9	100–790		DW	Williams et al. (2006)
	Leafy vegetables	West Bengal, India	Water, soil		<0.04–690		DW	Roychowdhury et al. (2002)
	Leafy vegetables	West Bengal, India	Water, soil		110–790	230	DW	Bhattacharya et al. (2010)
	Fruit vegetables	Bangladesh	Water, soil	64	50–1,590		DW	Williams et al. (2006)
	Fruit vegetables	West Bengal, India	Water, soil	65	0–360	140	DW	Bhattacharya et al. (2010)
	Fruit vegetables	Nepal	Water, soil	15	<0.01–140		DW	Dahal et al. (2008)
	Root/tuber vegetables	Bangladesh	Water, soil	18	<40–1,930		DW	Williams et al. (2006)

(continued)

Table 8.1 (continued)

Crop	Sample type	Country	As contamination	n	Range (μg kg^{-1})	Mean (μg kg^{-1})	Unit base	References
	Root/tuber vegetables	West Bengal, India	Water, soil	23	110–250	162	DW	Bhattacharya et al. (2010)
	Root/tuber vegetables	Nepal	Water, soil	15	<0.01–1,020		DW	Dahal et al. (2008)
	All types	Bangladesh	Water, soil	225	19–489		DW	Alam et al. (2003)
	All types	West Bengal, India	Water		34–167		DW	Signes-Pastor et al. (2008)
	All types	Bangladesh (sold in UK)		68	<5–540	55	DW	Al Rmalli et al. (2005)
	All types	Europe (sold in UK)		24	<5–87	24	DW	Al Rmalli et al. (2005)
	Fruit vegetables	Spain		15	20–84	7	DW	Matos-Reyes et al. (2010)
	Leaf vegetables	Spain		18	43–605		DW	Matos-Reyes et al. (2010)
	All types	Fujian, China		174	0.1–338	23	FW	Huang et al. (2006)
	All types	Canada		262	<0.1–84		FW	Dabeka et al. (1993)
	All types	USA			1.4–9.9		FW	Schoof et al. (1999)
	All types	England, UK		251	<0.5–25	3	FW	Weeks et al. (2007)
	Leafy vegetables	China		183	<1–223	10[c]	FW	Song et al. (2009)
	Fruit vegetables	China		152	<1–202	11[c]	FW	Song et al. (2009)
	Root vegetables	China		81	1–479	13[c]	FW	Song et al. (2009)
	All types	Chile (Santiago)		3		7	FW	Munoz et al. (2005)
	All type	Chile (Northern)			8–604		FW	Munoz et al. (2002)

Fruits	Papaya	Nepal	Water, soil	26	170–620	258	DW	Dahal et al. (2008)
	Papaya	West Bengal, India	Water, soil	3	156–373		DW	Roychowdhury et al. (2002)
	Banana	West Bengal, India	Water, soil	4	<0.04–54		DW	Roychowdhury et al. (2002)
	Lemon	West Bengal, India	Water, soil	30	0–50	12	DW	Bhattacharya et al. (2010)
	Various	US			1.6–40		FW	Schoof et al. (1999)
	Various	Canada		176	<0.1–37	4.5	FW	Dabeka et al. (1993)
	Various	Chile		3		7	FW	Munoz et al. (2005)

FW fresh weight, *DW* dry weight
[a]Twenty-six wheat cultivars grown on a single field site in one season
[b]Twenty-six wheat cultivars grown on a single field site in two seasons
[c]Geometrical mean

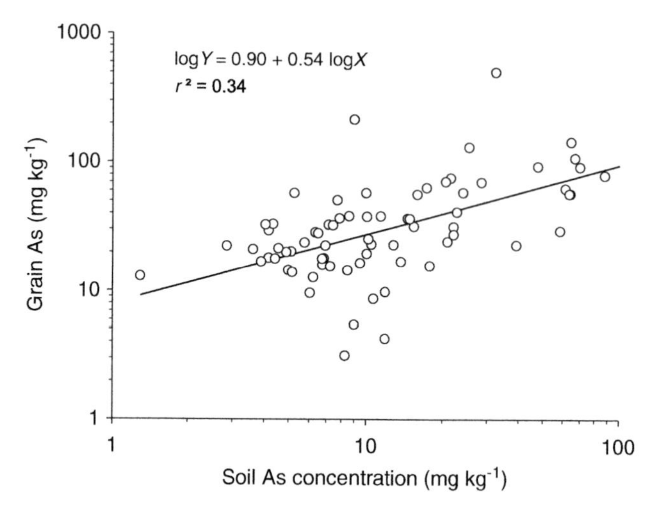

Fig. 8.1 Relationship between the arsenic concentration in wheat grain and total arsenic concentration in soil (Data are from Williams et al. (2007) and Zhao et al. (2010))

inorganic arsenic and methylated As, respectively, in a sample of wheat semolina produced in Spain. It is worth noting that Schoof et al. (1999) reported only 28% inorganic arsenic in wheat flour. However, this low percentage is likely to be an analytical artifact, because total and inorganic arsenic were determined separately using different procedures and the samples contained very low levels of arsenic bordering the detection limit. Compared with rice (see Chap. 2), the percentage of methylated arsenic species is much smaller in wheat. Plants are not thought to methylate arsenic in their tissues, so these species are produced in soils and taken up from soil solution (Chap. 6). The very low levels of organic arsenic species in wheat grain may reflect the aerobic soil environment under which wheat is grown. Arsenic methylation by soil microorganisms appears to be enhanced by the anaerobic conditions of flooded paddy soil (see Chap. 5).

The contribution to dietary intake of inorganic arsenic from wheat is generally small except in arsenic contaminated areas where locally produced wheat is consumed in large quantities. Using the mean values of arsenic in cereals excluding rice (15 and 40 μg kg^{-1} fresh weight for the lower and upper bound, respectively) reported by European Food Safety Authority (2009), assuming a daily consumption of 0.2 kg (fresh weight) wheat per person, which is typical for European adults (D'Amato et al. 2011), and 95% inorganic arsenic (Cubadda et al. 2010), the inorganic arsenic intake from wheat would amount to 2.8–7.6 μg $person^{-1}$ day^{-1}.

8.3 Maize

Maize (corn) is the number one crop in the world in terms of the total annual production (>800 million tonnes) (International Grains Council 2011). A number of papers have reported arsenic concentrations in maize grain (Table 8.1). Where the environment is not contaminated with arsenic, the mean concentration of arsenic in maize is generally within the range of 10–50 μg kg^{-1} dry weight, which is similar to that for wheat grain. Adomako et al. (2011) reported that the mean total arsenic concentration in Ghanaian maize, sorghum and millet samples (10 μg kg^{-1} dry weight) was an order of magnitude lower than in Ghanaian rice. In a field study on the effect of poultry manure which contains 4-hydroxy-3-nitrobenzenearsonic acid (Roxarsone) as a feed additive, Liebhardt (1976) found no significant increase in the arsenic concentration in maize grain. Elevated concentrations were reported when As-contaminated groundwater was used for irrigation in Pakistan (Baig et al. 2011). There are no reports of arsenic speciation in maize grain, although it is likely to be dominated by inorganic forms as has been shown in wheat. The contribution of maize to dietary intake of inorganic is likely to be small.

Earlier studies by Woolson et al. (1971, 1973) identified iron and aluminium oxides content of soil as a key factor affecting the toxicity of arsenic to maize. Interestingly, they found that phosphate additions to soil may either increase or decrease arsenic toxicity, depending on the interactions between Fe/Al oxides with P and As. The addition of phosphate to soil also increased arsenic uptake by maize plants, which was due to phosphate replacing adsorbed arsenate in soil (Woolson et al. 1973).

8.4 Potato

Potato tubers from uncontaminated soil generally contain <50 μg As kg^{-1} on the dry weight basis (Table 8.1; assuming a 20% dry matter content in fresh potato). However, arsenic contamination in soil or from irrigation water can markedly elevate the concentration of arsenic in potato tubers. For example, arsenic concentrations as high as 890 μg kg^{-1} (Williams et al. 2006), 1,020 μg kg^{-1} (Bhattacharya et al. 2010) and 1,390 μg kg^{-1} dry weight (Das et al. 2004) have been reported in potato tubers collected from areas of Bangladesh or West Bengal, India, which have been irrigated with As-contaminated groundwater. In an intensively cultivated agricultural area of central Spain, irrigation with As-contaminated water increased arsenic concentration potato tuber by 35 times (Moyano et al. 2009). Arsenic appears to accumulate preferentially on the skin of potato (Roychowdhury et al. 2002; Warren et al. 2003), either because tubers are able to absorb arsenic from the surrounding soil or the soil particles adhered on the tuber surface have not been completely cleaned.

Williams et al. (2006) determined arsenic speciation in a potato tuber sample from Bangladesh and found only inorganic arsenic, with no detectable organic arsenic species, whilst Signes-Pastor et al. (2008) reported the presence of MMA in potato tubers collected in a West Bengal village. For people consuming large quantities of potatoes grown in As-contaminated soils or irrigated with high-As groundwater, the intake of inorganic arsenic from this food source may be of concern.

8.5 Pulses

Pulses (legume seeds) are an important dietary source of proteins for populations in south Asia. Surveys in the As-impacted areas in West Bengal, India and Bangladesh showed relatively low concentrations of arsenic in legume seed, with mean values of 15–96 μg kg^{-1} dry weight (Table 8.1) (Bhattacharya et al. 2010; Roychowdhury et al. 2002; Williams et al. 2006). A similar range was found for pulses in Spain (Matos-Reyes et al. 2010). Only inorganic arsenic was found in five different pulses (Williams et al. 2006). The contribution from pulses to the dietary intake of arsenic is small because of the relatively low arsenic concentration and small quantity of pulses consumed (Williams et al. 2006).

8.6 Vegetables

A number of surveys have been conducted on the arsenic concentrations in various types of vegetables in the As-impacted areas of Bangladesh and West Bengal, India (Table 8.1) (Alam et al. 2003; Bhattacharya et al. 2010; Dahal et al. 2008; Roychowdhry 2008; Roychowdhury et al. 2002; Signes-Pastor et al. 2008; Smith et al. 2006; Williams et al. 2006). These surveys show a very wide range of values from below the detection limit to nearly 2,000 μg kg^{-1} dry weight. The mean concentration of arsenic in vegetables from these regions is generally within the range of 100–300 μg kg^{-1} dry weight, although in some cases the concentration can reach >1,000 μg kg^{-1} dry weight probably as a result of irrigation with high-As groundwater.

Warren et al. (2003) reported very high concentrations of arsenic (1–17.8 mg kg^{-1} dry weight) in lettuce grown on two contaminated soils (total As 65 and 748 mg kg^{-1}) in England. Vegetables collected from the vicinity of mining and smelting operations in Chenzhou, China, where the soils were heavily contaminated, were found to contain very high levels of arsenic: 2.1–54.5 mg kg^{-1} dry weight in cabbage, 1.5–4.7 mg kg^{-1} in eggplant fruits and 0.4–2.6 mg kg^{-1} in capsicum fruits (Liao et al. 2005). Leafy and root vegetables tend to accumulate more arsenic than fruit vegetables (Huang et al. 2006; Liao et al. 2005; Munoz et al. 2002), probably because the latter have a shoot to fruit translocation barrier. Growing leafy and root vegetables on As-contaminated soils should therefore be avoided.

Low concentrations of arsenic were found in vegetables in market basket surveys conducted in Canada and US (Dabeka et al. 1993; Schoof et al. 1999), Spain (Matos-Reyes et al. 2010), and from allotments in England (Weeks et al. 2007), reflecting the uncontaminated environment in which vegetables were grown, or a low bioavailability of arsenic in the allotment soils. Vegetables produced in Beijing, China and the surrounding areas contained <1–479 μg As kg^{-1} fresh weight, with a geometrical mean of 13 μg As kg^{-1} fresh weight (Table 8.1) (Song et al. 2009). Interestingly, vegetables produced in open fields were found to contain more arsenic than those produced inside greenhouses, suggesting that atmospheric deposition resulted in elevated arsenic concentrations in the outdoor vegetables (Song et al. 2009). Note that different studies report arsenic concentrations on either fresh weight or dry weight basis. Depending on the species of vegetable, the dry matter content can vary substantially, from 5% to 36% in the study of Williams et al. (2006) and from 4% to 15% in the study of Song et al. (2009).

With regard to arsenic speciation, Helgesen and Larsen (1998) found only inorganic arsenic in carrot samples grown in a contaminated soil with no evidence of the presence of methylated arsenic species. Similarly, Williams et al. (2006) found only inorganic arsenic in a number of vegetables collected in Bangladesh. Signes-Pastor et al. (2008) analysed arsenic speciation in number of vegetables, spices and rice collected in a village of West Bengal, India. They found inorganic arsenic to be the only arsenic species in kidney beans, tomato, onion, betel nut, cauliflower and brinjal. MMA was detected in carrot and radish, accounting for 20–39% of the total arsenic extracted. In contrast, some of the spices analysed had a higher percentage of MMA: 46% in coriander seeds, 66% in turmeric powder and 82% in fenugreek seeds. The high proportions of MMA in spice seeds may be related to the high translocation efficiency of methylated arsenic species in plants (see Chap. 6). No DMA was detected in vegetables or spices, which is different from rice which contained DMA but no MMA (Signes-Pastor et al. 2008). Roychowdhry (2008) reported an average of 89% inorganic arsenic and 11% DMA in leafy vegetables from As-affected areas in West Bengal, while Smith et al. (2006) obtained a mean of 96% inorganic arsenic in vegetables consumed by households in Munshiganj and Monohardi, Bangladesh. It can be concluded that, for most vegetables, inorganic arsenic is the predominant species of arsenic.

The contribution of vegetables to dietary arsenic intake depends on the amount and the types of vegetables consumed, and the concentrations of arsenic in different vegetables. For a rural population in West Bengal with an average daily consumption of ~380 g rice, 200 g vegetables (fresh weight) and 1.8 L of water containing 50 μg As L^{-1}, the intake of inorganic arsenic was 88, 62 and 1 μg $person^{-1}$ day^{-1} from water, rice and vegetables, respectively (Signes-Pastor et al. 2008); thus, vegetables contributed <1% to the total intake of inorganic arsenic in this scenario. However, Signes-Pastor et al. (2008) used a relatively low arsenic concentration in vegetables (6 μg kg^{-1} fresh weight) in their calculations, whereas other surveys in West Bengal and Bangladesh have shown that arsenic concentrations in vegetables can be considerably higher (Table 8.1). Other estimates of inorganic arsenic intake from vegetables for the rural population in the Bengal delta vary from 1 to 17 μg $person^{-1}$ day^{-1} for adults (Roychowdhry 2008; Roychowdhury et al. 2003; Williams et al. 2006); these values are still relatively small compared with the intakes from rice and drinking water. Alam et al. (2003) and Matos-Reyes et al. (2010) reported intake values of 27.8 and 31 μg $person^{-1}$ day^{-1}, respectively. However, their calculations were probably made incorrectly by multiplying the fresh weigh of vegetables consumed with the arsenic concentration based on dry weight; the actual intake of inorganic arsenic from vegetables was likely to be around tenfold lower. For the residents in Beijing, China, the average intake of arsenic from vegetables was about 5 μg day^{-1} for adults (Song et al. 2009). For the population in Santiago, Chile, an average intake of 2.3 μg As $person^{-1}$ day^{-1} from vegetables was estimated (Munoz et al. 2005).

8.7 Fruits

The information of arsenic concentrations and speciation in fruits is more limited than in other food categories (Table 8.1). Based on the surveys by Dahal et al. (2008) and Roychowdhury et al. (2002), papaya produced in West Bengal, India and Nepal, both affected by arsenic contaminated irrigation water, contains considerable levels of As, whereas banana and lemon have relatively low concentrations of arsenic (Bhattacharya et al. 2010; Roychowdhury et al. 2002). In market basket surveys conducted in Canada, US and Santiago, Chile, fruits generally contain low levels of arsenic (Dabeka et al. 1993; Munoz et al. 2005; Schoof et al. 1999).

In terms of arsenic speciation, there have been no studies that directly determine the presence or absence of different arsenic species in fruits. Schoof et al. (1999) reported both total and inorganic arsenic in different foods, including fruits, in a market basket survey. In their dataset inorganic arsenic accounted for 21–150% of the total As. However, as discussed above (see Sect. 8.2), the difference between total and inorganic arsenic may be attributed to different analytical methods used rather than the presence of organic arsenic species. In general, the contribution of fruits to dietary arsenic intake is likely to be very small (Yost et al. 1998); an average intake of 1.1 μg As $person^{-1}$ day^{-1} from fruits was estimated for the population in Santiago, Chile (Munoz et al. 2005).

References

Adomako EA, Williams PN, Deacon C, Meharg AA (2011) Inorganic arsenic and trace elements in Ghanaian grain staples. Environ Pollut 159:2435–2442

Al Rmalli SW, Haris PI, Harrington CF, Ayub M (2005) A survey of arsenic in foodstuffs on sale in the United Kingdom and imported from Bangladesh. Sci Total Environ 337:23–30

Alam MGM, Snow ET, Tanaka A (2003) Arsenic and heavy metal contamination of vegetables grown in Samta village, Bangladesh. Sci Total Environ 308:83–96

Baig JA, Kazi TG, Shah AQ, Afridi HI, Kandhro GA, Khan S, Kolachi NF, Wadhwa SK, Shah F, Arain MB, Jamali MK (2011) Evaluation of arsenic levels in grain crops samples, irrigated by tube well and canal water. Food Chem Toxicol 49:265–270

Bhattacharya P, Samal AC, Majumdar J, Santra SC (2010) Arsenic contamination in rice, wheat, pulses, and vegetables: a study in an arsenic affected area of West Bengal, India. Water Air Soil Pollut 213:3–13

Cubadda F, Ciardullo S, D'Amato M, Raggi A, Aureli F, Carcea M (2010) Arsenic contamination of the environment-food chain: a survey on wheat as a test plant to investigate phytoavailable arsenic in Italian agricultural soils and as a source of inorganic arsenic in the diet. J Agric Food Chem 58:10176–10183

D'Amato M, Aureli F, Ciardullo S, Raggi A, Cubadda F (2011) Arsenic speciation in wheat and wheat products using ultrasound- and microwave-assisted extraction and anion exchange chromatography-inductively coupled plasma mass spectrometry. J Anal Atom Spectrom 26:207–213

Dabeka RW, McKenzie AD, Lacroix GMA, Cleroux C, Bowe S, Graham RA, Conacher HBS, Verdier P (1993) Survey of arsenic in total diet food composites and estimation of the dietary intake of arsenic by Canadian adults and children. J AOAC Int 76:14–25

Dahal BM, Fuerhacker M, Mentler A, Karki KB, Shrestha RR, Blum WEH (2008) Arsenic contamination of soils and agricultural plants through irrigation water in Nepal. Environ Pollut 155:157–163

Das HK, Mitra AK, Sengupta PK, Hossain A, Islam F, Rabbani GH (2004) Arsenic concentrations in rice, vegetables, a fish in Bangladesh: a preliminary study. Environ Int 30:383–387

Diaz OP, Leyton I, Munoz O, Nunez N, Devesa V, Suner MA, Velez D, Montoro R (2004) Contribution of water, bread, and vegetables (raw and cooked) to dietary intake of inorganic arsenic in a rural village of Northern Chile. J Agric Food Chem 52:1773–1779

European Food Safety Authority (2009) Scientific opinion on arsenic in food. EFSA J 7:1351

Helgesen H, Larsen EH (1998) Bioavailability and speciation of arsenic in carrots grown in contaminated soil. Analyst 123:791–796

Huang RQ, Gao SF, Wang WL, Staunton S, Wang G (2006) Soil arsenic availability and the transfer of soil arsenic to crops in suburban areas in Fujian Province, southeast China. Sci Total Environ 368:531–541

International Grains Council (2011) Grains market report, GMR No. 414. International Grains Council, London

Jelinek CF, Corneliussen PE (1977) Levels of arsenic in United States food supply. Environ Health Perspect 19:83–87

Liao XY, Chen TB, Xie H, Liu YR (2005) Soil As contamination and its risk assessment in areas near the industrial districts of Chenzhou City, Southern China. Environ Int 31:791–798

Liebhardt WC (1976) Arsenic content of corn grain grown on a coastal plain soil amended with poultry manure. Commun Soil Sci Plant Anal 7:169–174

Matos-Reyes MN, Cervera ML, Campos RC, de la Guardia M (2010) Total content of As, Sb, Se, Te and Bi in Spanish vegetables, cereals and pulses and estimation of the contribution of these foods to the Mediterranean daily intake of trace elements. Food Chem 122:188–194

Moyano A, Garcia-Sanchez A, Mayorga P, Anawar HM, Alvarez-Ayuso E (2009) Impact of irrigation with arsenic-rich groundwater on soils and crops. J Environ Monit 11:498–502

Munoz O, Diaz OP, Leyton I, Nunez N, Devesa V, Suner MA, Velez D, Montoro R (2002) Vegetables collected in the cultivated Andean area of northern Chile: total and inorganic arsenic contents in raw vegetables. J Agric Food Chem 50:642–647

Munoz O, Bastias JM, Araya M, Morales A, Orellana C, Rebolledo R, Velez D (2005) Estimation of the dietary intake of cadmium, lead, mercury, and arsenic by the population of Santiago (Chile) using a Total Diet Study. Food Chem Toxicol 43:1647–1655

Norra S, Berner ZA, Agarwala P, Wagner F, Chandrasekharam D, Stuben D (2005) Impact of irrigation with As rich groundwater on soil and crops: a geochemical case study in West Bengal delta plain, India. Appl Geochem 20:1890–1906

Reyes MNM, Cervera ML, Campos RC, de la Guardia M (2007) Determination of arsenite, arsenate, monomethylarsonic acid and dimethylarsinic acid in cereals by hydride generation atomic fluorescence spectrometry. Spectrochim Acta B Atom Spectrosc 62:1078–1082

Roychowdhry T (2008) Impact of sedimentary arsenic through irrigated groundwater on soil, plant, crops and human continuum from Bengal delta: special reference to raw and cooked rice. Food Chem Toxicol 46:2856–2864

Roychowdhury T, Uchino T, Tokunaga H, Ando M (2002) Survey of arsenic in food composites from an arsenic-affected area of West Bengal, India. Food Chem Toxicol 40:1611–1621

Roychowdhury T, Tokunaga H, Ando M (2003) Survey of arsenic and other heavy metals in food composites and drinking water and estimation of dietary intake by the villagers from an arsenic-affected area of West Bengal, India. Sci Total Environ 308:15–35

Schoof RA, Yost LJ, Eickhoff J, Crecelius EA, Cragin DW, Meacher DM, Menzel DB (1999) A market basket survey of inorganic arsenic in food. Food Chem Toxicol 37:839–846
Shewry PR (2009) Wheat. J Exp Bot 60:1537–1553
Signes-Pastor AJ, Mitra K, Sarkhel S, Hobbes M, Burlo F, De Groot WT, Carbonell-Barrachina AA (2008) Arsenic speciation in food and estimation of the dietary intake of inorganic arsenic in a rural village of West Bengal, India. J Agric Food Chem 56:9469–9474
Skrbic B, Cupic S (2005) Toxic and essential elements in soft wheat grain cultivated in Serbia. Eur Food Res Technol 221:361–366
Smith NM, Lee R, Heitkemper DT, Cafferky KD, Haque A, Henderson AK (2006) Inorganic arsenic in cooked rice and vegetables from Bangladeshi households. Sci Total Environ 370:294–301
Song B, Lei M, Chen TB, Zheng YM, Xie YF, Li XY, Gao D (2009) Assessing the health risk of heavy metals in vegetables to the general population in Beijing, China. J Environ Sci 21:1702–1709
Su YH, McGrath SP, Zhao FJ (2010) Rice is more efficient in arsenite uptake and translocation than wheat and barley. Plant Soil 328:27–34
Warren GP, Alloway BJ, Lepp NW, Singh B, Bochereau FJM, Penny C (2003) Field trials to assess the uptake of arsenic by vegetables from contaminated soils and soil remediation with iron oxides. Sci Total Environ 311:19–33
Weeks CA, Brown SN, Vazquez I, Thomas K, Baxter M, Warriss PD, Knowles TG (2007) Multi-element survey of allotment produce and soil in the UK. Food Addit Contam 24:877–885
Wiersma D, van Goor BJ, van der Veen NG (1986) Cadmium, lead, mercury, and arsenic concentrations in crops and corresponding soils in the Netherlands. J Agric Food Chem 34:1067–1074
Williams PN, Islam MR, Adomako EE, Raab A, Hossain SA, Zhu YG, Feldmann J, Meharg AA (2006) Increase in rice grain arsenic for regions of Bangladesh irrigating paddies with elevated arsenic in groundwaters. Environ Sci Technol 40:4903–4908
Williams PN, Villada A, Deacon C, Raab A, Figuerola J, Green AJ, Feldmann J, Meharg AA (2007) Greatly enhanced arsenic shoot assimilation in rice leads to elevated grain levels compared to wheat and barley. Environ Sci Technol 41:6854–6859
Woolson EA, Axley JH, Kearney PC (1971) Correlation between available soil arsenic, estimated by 6 methods, and response of corn (*Zea mays* L.). Soil Sci Soc Am Proc 35:101–105
Woolson EA, Axley JH, Kearney PC (1973) Chemistry and phytotoxicity of arsenic in soils. 2. Effects of time and phosphorus. Soil Sci Soc Am J 37:254–259
Yost LJ, Schoof RA, Aucoin R (1998) Intake of inorganic arsenic in the North American diet. Hum Ecol Risk Assess 4:137–152
Zhao FJ, Stroud JL, Eagling T, Dunham SJ, McGrath SP, Shewry PR (2010) Accumulation, distribution, and speciation of arsenic in wheat grain. Environ Sci Technol 44:5464–5468

Index

A.A. Meharg and F.J. Zhao, *Arsenic & Rice*, DOI 10.1007/978-94-007-2947-6,